Functions and Graphs

A Clever Study Guide

AMS/MAA | PROBLEM BOOKS

VOL **29**

Functions and Graphs

A Clever Study Guide

James Tanton

MAA PRESS

An Imprint of the AMERICAN MATHEMATICAL SOCIETY

Providence, Rhode Island

2010 *Mathematics Subject Classification*. Primary 97-XX.

For additional information and updates on this book, visit
www.ams.org/bookpages/prb-29

Library of Congress Cataloging-in-Publication Data

Cataloging-in-Publication Data has been applied for by the AMS.
See http://www.loc.gov/publish/cip/.

Contents

About These Study Guides

The Mathematical Association of America's American Mathematics Competitions' website, www.maa.org/math-competitions, announces loud and clear:

> Teachers and schools can benefit from the chance to challenge students with interesting mathematical questions that are aligned with curriculum standards at all levels of difficulty.

For over six decades the dedicated and clever folks of the MAA have been creating and collating marvelous, stand-alone mathematical tidbits and sharing them with the world of students and teachers through mathematics competitions. Each question serves as a portal for deep mathematical mulling and exploration. Each is an invitation to revel in the mathematical experience.

And more! In bringing together all the questions that link to one topic, a coherent mathematical landscape, ripe for a guided journey of study, emerges. The goal of this series is to showcase the landscapes that lie within the MAA's competition resources and to invite students, teachers, and all life-long learners, to engage in the mathematical explorations they invite. Learners will not only deepen their understanding of curriculum topics but also gain the confidence to play with ideas and work to become agile intellectual thinkers.

I was recently asked by some fellow mathematics educators what my greatest wish is for our next generation of students. I responded:

> ...a personal sense of curiosity coupled with the confidence to wonder, explore, try, get it wrong, flail, go on tangents, make connections, be flummoxed, try some more, wait for epiphanies, lay groundwork for epiphanies, go down false leads, find moments of success nonetheless, savor the "ahas", revel in success, and yearn for more.

Our complex society demands of our next generation not only mastery of quantitative skills, but also the confidence to ask new questions, to innovate, and to succeed. Innovation comes only from bending and pushing ideas and being willing to flail. One must rely on one's wits and on one's common sense. And one must persevere. Relying on memorized answers to previously asked—and answered!—questions does not push the frontiers of business research and science research.

The MAA competition resources provide today's mathematics thinkers, teachers, and doers

- the opportunity to learn and to teach problem solving and
- the opportunity to review the curriculum from the perspective of understanding and clever thinking, letting go of memorization and rote doing.

Each of these study guides

- runs through the entire standard curriculum content of a particular mathematics topic from a sophisticated and mathematically honest point of view,
- illustrates in concrete ways how to implement problem-solving strategies for problems related to the particular mathematics topic, and
- provides a slew of practice problems from the MAA competition resources along with their solutions.

As such, these guides invite you to

- review and deeply understand mathematics topics,
- practice problem solving,
- gain incredible intellectual confidence,

and, above all,

- to enjoy mathematics!

This Guide and Mathematics Competitions

Whether you enjoy the competition experience and are motivated and delighted by it or you, like me, tend to shy away from it, this guide is for you!

We all have our different styles and proclivities for mathematics thinking, doing, and sharing, and they are all good. The point, in the end, lies with the enjoyment of the mathematics itself. Whether you like to solve problems under the time pressure of a clock or while mulling on a stroll, problem solving is a valuable art that will serve you well in all aspects of life.

This guide teaches how to think about content and how to solve challenges. It serves both the competition doers and the competition nondoers. That is, it serves the budding and growing mathematicians we all are.

On Competition Names

This guide pulls together problems from the history of the MAA's American competition resources.

The competitions began in 1950 with the Metropolitan New York Section of the MAA offering a "Mathematical Contest" each year for regional high school students. The competitions became national endeavors in 1957 and adopted the name "Annual High School Mathematics Examination" in 1959. This was changed to the "American High School Mathematics Examination" in 1983.

In these guides, the code "#22, AHSME, 1972", for example, refers to problem number 22 from the 1972 AHSME, Annual/American High School Mathematics Examination.

In 1985 a contest for middle school students was created, the "American Junior High School Mathematics Examination", and shortly thereafter the contests collectively became known as the "American Mathematics Competitions", the AMC for short. In the year 2000 competitions limited to high school students in grades 10 and below were created and the different levels of competitions were renamed the AMC 8, the AMC 10, and the AMC 12.

In these guides, "#13, AMC 12, 2000", for instance, refers to problem number 13 from the 2000 AMC 12 examination.

In 2002, and ever since, two versions of the AMC 10 and the AMC 12 have been administered, about two weeks apart, and these are referred to as the AMC 10A, AMC 10B, AMC 12A, and AMC 12B.

In these guides, "#24, AMC 10A, 2013", for instance, refers to problem number 24 from the 2013 AMC 10A examination.

On Competition Success

Let's be clear:

> "I am using this guide for competition practice. Does this guide promise me 100% success on all mathematics competitions, each and every time?"

Of course not! But this guide does offer, if worked through with care,

- feelings of increased confidence when taking part in competitions,
- clear improvement on how you might handle competition problems,
- clear improvement on how you might handle your emotional reactions to particularly outlandish-looking competition problems.

Mathematics is an intensely human enterprise and one cannot underestimate the effect of emotions when doing mathematics and attempting to solve challenges. This guide gives the human story that lies behind the mathematics content and discusses the human reactions to problem solving.

As we shall learn, the first and the most important effective step in solving a posed problem is:

> **Step 1**: Read the question, have an emotional reaction to it, take a deep breath, and then reread the question.

This guide provides practical content knowledge, problem-solving tools and techniques, and concrete discussion on getting over the barriers of emotional blocks. Even though its goal is not necessarily to improve competition scores, these are the tools that nonetheless lead to that outcome!

This Guide and the Craft of Solving Problems

Success in mathematics—however you wish to define it—comes from a strong sense of self-confidence: the confidence to acknowledge one's emotions and to calm them down, the confidence to pause over ideas and come to educated guesses or conclusions, the confidence to rely on one's wits to navigate through unfamiliar terrain, the confidence to choose understanding over impulsive rote doing, and the confidence to persevere.

Success and joy in science, business, and life doesn't come from programmed responses to preset situations. It comes from agile and adaptive thinking coupled with reflection, assessment, and further adaptation.

Students—and adults too—are often under the impression that one should simply be able to leap into a mathematics challenge and make instant progress of some kind. This is not how mathematics works! It is okay to fumble and flail and to try out ideas that turn out not to help in the end. In fact, this *is* the problem-solving process and making multiple false starts should not at all be dismissed! (Think of how we solve problems in everyday life.)

It is also a natural part of the problem-solving process to react to a problem.

> "This looks scary."
>
> "This looks fun."
>
> "I don't have a clue what the question is even asking!"
>
> "Wow. Weird! Could that really be true?"
>
> "Who cares?"
>
> "I don't get it."
>
> "Is this too easy? I am suspicious."

We are all human, and the first step to solving a problem is to come to terms with our emotional reaction to it—especially if that reaction is one of being overwhelmed. Step 1 to problem solving mentioned earlier is vital.

Once we have our nerves in check, at least to some degree, there are a number of techniques we could try in order to make some progress with the problem.

The ten strategies we briefly outline in Appendix I are discussed in full detail on the MAA's *Curriculum Inspirations* webpage, `www.maa.org/ci`. There you will find essays and videos explaining each technique in full, with worked out examples and slews of further practice examples and their solutions.

This guide also contains worked out examples. Look for the Featured Problems in select chapters where I share with you my own personal thoughts, emotions, and eventual approach in solving a given problem using one of the ten problem-solving strategies.

This Guide and Mathematics Content: Functions and Graphs

This guide covers the mathematical exploration of functions and their graphs. It is a swift overview, but it is complete in the context of the content discussed in beginning and advanced high school courses. The purpose of this book is to supplement and put into perspective the material of any course on the subject you may have taken or are currently taking. (This book will be tough going for those encountering this work for the very first time!)

In reading and working through the material presented here you will

- see the story of functions, first in a nonnumerical context, and then with a focus on numerical examples,

- make surprising connections between topics: sequences and functions; additive thinking and multiplicative reasoning; scatter plots and graphs of equations functions, to name a few pairings,

- begin to move away from memorization and half-understanding to deep understanding,

and thereby

- be equipped for agile, clever thinking in the subject.

This book will guide you to sound mathematical work on the topic of functions and, of course, to sound problem-solving skills as well.

Some Unorthodox "Gems"

There are five treatments of content in this volume that are sometimes considered unorthodox. They are:

- Chapter 5: Viewing all graphs as graphs of scatter plots.
- Chapter 6: Asking, in the transformation of functions, "which input is behaving like 0?"

- Chapter 8: Simplifying the algebra and graphing of quadratics by utilizing the power of symmetry.
- Chapters 8, 9, and 10: Focusing on "interesting x-values" to simplify graphing.
- Chapter 12: Obviating hard work in fitting functions to data.

You might want to look at these chapters early on.

Part I
Functions and Graphs

1
What Is a Function? A Swift Conceptual Overview

In the 1600s and 1700s mathematicians formalized the idea of transforming or performing actions on numbers or on points in the plane or, more generally, on elements of arbitrary sets of objects. The idea is to simply regard a transformation as a rule for matching each element of one set with another element—its transform—of a second set.

Around 1675, calculus scholar Gottfried Wilhelm Leibniz coined the word "function" for these rules basing the name on the Latin word *fungi* meaning "to perform". (This is not to be confused with the Greek root word for "sponge" that led to the modern botanical word *fungus*.) This thinking and work was popularized by Leonhard Euler, who is credited as the first to use the notation "$f(x)$" associated with functions so familiar to us today.

Here is a swift summary of the key ideas and notation of modern-day function work. In school textbooks, one might be given the impression that functions are always numerical. This is not at all the case. In this chapter we'll show that nonnumerical examples are actually easier to think about and serve well for making function concepts straightforward and clear. We'll focus on numerical functions in later chapters.

Let's get started.

Suppose we have two sets of objects X and Y. A *function* from X to Y is a rule, supposedly clearly understood by all, which assigns to each and every object from the set X precisely one object from the set Y.

Example. If X is the set of American citizens and Y is the set of nine-digit numbers, then the rule that assigns to each citizen a Social Security number is a function.

Example. Let X be the set of points in the plane and let Y be the same. Then a given rotation is a function: it assigns to each point its image point.

If we use the letter f, say, to denote the function, then we write

$$f : X \to Y.$$

We usually call each element x of X an *input* of the function and denote the element assigned to it from Y by $f(x)$. This is called the matching *output* for the input x.

Comment. $f(x)$ is read out loud as "f of x".

Example. Consider the rule, which we shall denote by M, that assigns to each person his or her biological mother. This is a well-defined rule from the set of all people of the world to the set of all women, and so we have a function.

$$M : \text{People} \to \text{Women}.$$

My name is James and mother's name is Abby and so in the context of these two specific people we have

$$M(\text{James}) = \text{Abby},$$

which reads "the mother of James is Abby".

(There are many people named James and Abby in the world. In working with functions each possible input and each possible output should have a unique identifier. First names, alas, are usually not unique to people. But let's work here in the context of my immediate family, where first names do happen to be unique.)

Thinking of more people in my family, we have $M(\text{Turner}) = \text{Lindy}$ and $M(\text{Lindy}) = \text{Sally}$. We also have $M(M(\text{Turner})) = \text{Sally}$ to be read as "the mother of the mother of Turner is Sally".

This rule is indeed a function: each and every person is "assigned" one, and only one, biological mother. That two or more people can be assigned the same mother (siblings!) does not violate the definition of being a function, nor does the fact that not all women are mothers.

Example. Consider the rule

$$A : \text{People} \to \text{Natural Numbers}$$

that assigns to each living person his or her age expressed as a whole number of years.

This is a function, and at the time of writing this chapter we have $A(\text{James}) = 50$. That there are many people the same age as I does not violate the definition of this rule being a function.

Example. Consider the "rule"

$$D : \{\text{Men over the age of }40\} \to \text{People}$$

which assigns to each man over the age of 40 his daughter.

This is not a function as the rule has two problems: not every man has a daughter, and some men have more than one daughter (which do we assign?). We say that this rule is not *well-defined*.

We could salvage the previous example by restricting D to the set of all men over the age of 40 who have at least one daughter and assign to each such fellow his youngest daughter.

Example. Consider the truth function

$$T : \text{The set of all possible statements} \to \{\text{True, False}\}$$

given by the rules:

Assign the word "true" to a statement if the statement is true.

Assign the word "false" to a statement if the statement is false.

For example,

$$T(\text{Liquid water is wet.}) = \text{True.}$$
$$T(\text{James likes cooked green peppers.}) = \text{False.}$$
$$T(\text{This sentence is true.}) = ?$$

This example shows that our loose definition of a function as a rule is problematic. What constitutes a meaningful rule?

An Attempt at a Formal Definition of a Function

Rather than think of a function as defined as a rule, one can instead, in an attempt to be formal, define a function to simply be a set, a set of special pairs constructed from two given sets. In this—somewhat dry— thinking, a function f from a set X to a set Y is a collection of ordered pairs (x, y), with x an element of X and y an element of Y, with the property that for each x in X there is precisely one pair in the collection f of pairs with first element x.

For example, if $X = \{1, 2, 3\}$ and $Y = \{A, B\}$, then the collection of pairs

$$f = \{(1, A), (2, C), (3, A)\}$$

is a function, but the collections

$$g = \{(1, A), (1, B), (2, A), (3, C)\}$$

and

$$h = \{(2, B), (3, A)\}$$

are not functions from X to Y. (The first assigns more than one element of Y to the element 1 of X, while the seconds fails to assign any element to it.)

Comment. One can attempt to be even more formal, and hence dryer and less intuitive, and say:

> For a pair of sets X and Y, let $X \times Y$ denote the set of all possible ordered pairs (x, y) with x from X and y from Y. Then a function f from X to Y is a subset of $X \times Y$ with the property that each and every element x of X appears as a first element of exactly one pair in f.

To mathematicians being formal, a function is then just a special subset of $X \times Y$. This gives the feel of being clear and precise, but it is actually just as problematic as our loose opening definition: how do you describe which pairs (x, y) belong to the special subset? For example, if X is the set of all the people of the world and Y is the set of all men, then one still needs to say something like "(x, y) belongs to the father function only if y is x's biological father".

Exercise. If X and Y are sets, is it possible for the entire set of pairs $X \times Y$ to be a function?

Solution. Yes, if Y is a set with only one element. $\square$

Exercise. A function assigns to each element x of a set X the element x. (This is called the *identity* function. It does nothing to inputs.) Describe the subset of $X \times X$ that defines this function.

Solution. We have the "diagonal subset", the set of all pairs of the form (x, x) with x in X. $\square$

Exercise. If A is a set with a elements in it and B is a set with b elements, how many different functions $f : A \to B$ are there?

Solution. For each element in the set A we have b choices of which element of B to assign to it. There are thus $b \times b \times \cdots \times b = b^a$ possible functions. $\qquad\qquad\square$

Some Jargon

The set of all allowable inputs of a function is called the *domain* of the function. The set of all possible outputs is called its *range*.

Comment. If we write $f : X \to Y$, then it is implied that the domain of the function is X. The range of the function, however, need not be all of Y. For example, for the age function described above

$$A : \text{People} \to \text{Natural Numbers}$$

the domain of the function is the set of all people of the world and the range is the set of whole numbers $\{0, 1, 2, 3, 120(?)\}$. (How old is the oldest current living person?)

Example. Consider the function

$$F : \text{Counting Numbers} \to \text{Counting Numbers}$$

which assigns to each counting number its first digit. (So $F(902) = 9$ and $F(8) = 8$.) Here the domain is the set of all counting numbers and the range is $\{1, 2, 3, 4, 5, 6, 7, 8, 9\}$.

Compostion of Functions

One can imagine a function to be a machine that takes inputs and churns each into an output. (See Figure 1.) As such, we can imagine linking together two, or more, machines.

Figure 2 assumes that each output of a function $f : X \to Y$ is a valid input of a function $g : Y \to Z$. The action of these two functions together take an element x of X and first hits it with f to obtain $f(x)$ and then "hits" this output with g to give $g(f(x))$.

Example. Consider the father function F that assigns to each living person his or her biological father and the year of birth function Y that

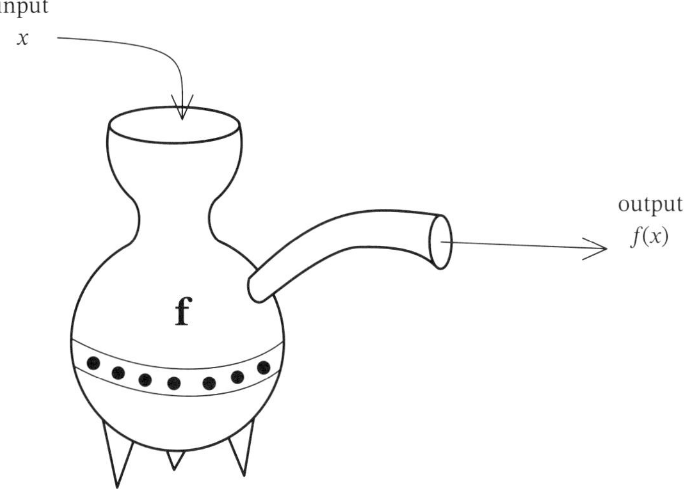

FIGURE 1

assigns to each person the year he or she was born. Then $Y(F(\text{James})) =$ 1946, the year my father was born. And $F(Y(\text{James}))$ makes no sense as the year I was born has no biological father.

Comment. The notation we use for functions can be confusing here. If we write $h(a(w))$, for instance, then we must assume that w is a general symbol for an input and that h and a are the names of functions. The

FIGURE 2

order in which these functions were applied is reverse to what many expect: first a was applied to w and then h was applied to $a(w)$ (and we must also assume that w is a valid input for the function a).

By the way, this way of reading $h(a(w))$ matches society's general rule for dealing with nested parentheses: focus on the innermost parentheses first and move outwards from there.

Exercise. With the function machines f and g shown in Figure 2, draw a picture of a linkage of machines that takes an input x and eventually produces $f(f(g(f(g(x)))))$.

Solution. Draw a g machine on the left with output hose going into an f machine with output hose into a g machine with output hose going into an f machine with output hose going into another f machine. $\square$

Given a function $f : X \to Y$ and a function $g : Y \to Z$, the function that assigns an element x of X the element $g(f(x))$ of Z is called the *composition* of the functions f and g.

Just to make matters particularly confusing, some people like to denote the composition of f and g as $g \circ f$, which is to be read backwards: the function $g \circ f$ first applies f to an input and then applies g to the result.

People might also write $(g \circ f)(x) = g(f(x))$ or sometimes just $g \circ f(x) = g(f(x))$, which are each probably more confusing than helpful.

Exercise. Can you unravel what $(g \circ f)(x) = g(f(x))$ is actually saying?

Solution. It says: "The function with the strange name $g \circ f$ assigns to each input x the output $g(f(x))$." $\square$

The notation $g \circ f$ is read as "g composed with f".

If F is the father function and Y is the year of birth function described earlier, then $Y \circ F$ is meaningful (it is the function that assigns to each person the year his or her biological father was born). Also $Y \circ F \circ F$ is meaningful. (It is the function that assigns to each person the year his or her paternal grandfather was born.) The composition $F \circ Y$ is not meaningful.

Sometimes it is helpful to read the little "$\circ$" for the composition of functions out loud also as "of". For instance, the function $Y \circ F$ is the "year of birth of the father of" function, and $Y \circ F \circ F$ is the "year of birth of the father of the father of…" function, and so on. Reading $F \circ Y$ out loud ("the father of the year of birth of…") makes it clear that this composition is meaningless.

Comment. Even though using the word "of" forces us to read $g \circ f$ from left to right, we must remember that it is the rightmost function f that is applied first to an input x and then the function g. This mismatch of direction is a result of our initial choice of notation for a function output. Life would be considerably easier if we wrote $(x)f$, or even just xf, for the result of applying the function f to an input x. Then:

xf reads: start with x and then apply f to it,

xfg reads: start with x and then apply f to it and then apply g to the result,

and so on. All would be consistently left to right. If x represents me, James, and f is the father function, then the $f(x)$ notation follows the language "the father of James". The xf notation follows the language "James's father". The mathematics community has settled on the first style.

Comment. In the early grades students are taught to write the product of two numbers three ways. For example, two times three could be written with a cross symbol, 2×3; or with a raised dot, $2 \cdot 3$; or with parentheses, $2(3)$. This third way matches function notation, which adds yet another layer of possible confusion. (Many students, when they first see the notation $f(x)$, naturally think $f \times x$.) It would be deliciously confusing if I called the function that adds two to each and every number "2". In this context, we'd have $2(3) = 5$. As is always the case in reading mathematics, context is important.

Exercise. Let

$$M : \text{People} \to \text{People}$$

be the function that assigns to each person his or her biological mother, let

$$F : \text{People} \to \text{People}$$

be the function that assigns to each person his or her biological father, let

$$A : \text{People} \to \text{Whole Numbers}$$

be the function that assigns to each person his or her age in years, and let

$$L : \text{People} \to \text{Letters}$$

be the function that assigns to each person the first letter of his or her name.

Which of the following compositions of functions are meaningful? For those that make sense, describe what the composite function is doing.

    a) $L \circ M \circ F \circ F$
    b) $A \circ L \circ L$
    c) $L \circ A \circ F$
    d) $M \circ L \circ F$

Solution. Only the first is meaningful. It gives the first letter of the name of a person's great-grandmother. $\square$

Iterated Functions

One can sometimes compose a function with itself to obtain an *iterated* function.

For example, if M is the function that assigns to each person his or her biological mother, then $M \circ M$ assigns to each person his or her maternal grandmother, $M \circ M \circ M$ his or her maternal great-grandmother, and so on.

If $S : \text{Counting Numbers} \to \text{Counting Numbers}$ is the successor function, it assigns to each counting number a the next counting number $a + 1$, so

$$S(a) = a + 1,$$
$$S(S(a)) = a + 2,$$
$$S(S(S(a))) = a + 3.$$

To iterate a function, each output of the function must be a valid input for the function.

Writing lists of compositions and writing lists of nested parentheses soon becomes tiresome. If f is a function that can be iterated, then, for a counting number k, we write f^k for that function composed with itself k times.

$$f^k(x) = \overset{k \text{ times}}{f \circ f \circ \cdots \circ f}(x) = f(f(\cdots f(x))).$$

We have the function machine picture of Figure 3. For example, $M^2(\text{James})$ is James's maternal grandmother and $S^n(a) = a + n$.

FIGURE 3

Exercise. If a and b are counting numbers, why is $f^a \circ f^b = f^{a+b}$?

Solution. Applying a function f to an input first b times and then a more times has the same effect as applying the function f to that input $a + b$ times. □

Comment. Annoyingly, this composition notation for iteration is abandoned when it comes to trigonometric functions. For example, $\sin^2(x)$ is taken to mean $(\sin(x))^2 = \sin(x) \times \sin(x)$ and not the composition $\sin(\sin(x))$. Annoying indeed!

Going Quirky

Assume $f : X \to X$ is a function that can be iterated.

Suppose we run an input x through the function machine f zero times; that is, we do nothing with the input. Then we are left holding the same value x. For this reason, it seems appropriate to declare

$$f^0(x) = x,$$

that is, to declare f^0 to be the *identity* function (assign to each input itself).

Can we give meaning to f^{-1}? Does it make sense to take a value x and run it through the machine -1 times? Perhaps this means we are attempting to run a value x backwards through the machine. (See Figure 4.)

FIGURE 4

To make sense of this, we must first assume that the value of x we are working with is in the range of f. Then, in thinking of the backwards operation of the machine, we see that we are looking for an input a whose output is x. Then $f^{-1}(x)$ would be a. That is,

$f^{-1}(x)$ is the input to f that gives the output x.

This, of course, is assuming that such an input exists and is clearly defined. Depending on the function, this might or might not be the case.

Example. Consider the function

$$R : \text{Strings of Letters} \to \text{Strings of Letters}$$

which assigns to each string of letters the string with the letters reversed. For example,

$$R(abc) = cba, \quad R(mmmmm) = mmmmm, \quad \text{and} \quad R(w) = w.$$

Each string of letters is an output of this function and each output comes from a unique input. Thus R^{-1} is meaningful, and we have, for instance, $R^{-1}(cba) = abc$ and $R^{-1}(w) = w$. (In fact, R^{-1} is identical to the function R.)

Example. Consider the function

$$F : \text{Men who are fathers} \to \text{Men who are fathers}$$

which assigns to each father his biological father. Then F^{-1} attempts to assign to each father his son. This is problematic as not all fathers have a son and, of those who do, some have more than one son and it is not clear which son to assign.

Exercise. Can this example be salvaged to some degree? Consider

$$F : \text{Males} \to \text{Men who have at least one son}$$

which assigns to each male his biological father. Set

$$S : \text{Men who have at least one son} \to \text{Males}$$

to be the function that assigns to all fathers of at least one son his oldest son. (We have a mismatch of domain and range now.) Is $F \circ S$ the identity function? Is $S \circ F$ the identity function?

Solution. The first is the identity function; the second is not. (Consider two brothers with $S \circ F$ applied to the younger brother.) $\square$

If $f : X \to X$ is a function such that for each x in X there is a unique element a in X which has output x (see Figure 5), then f^{-1} is defined to be the function that assigns to x the input from which it came: $f^{-1}(x) = a$. The function f^{-1} is called the *inverse* function to f.

Comment. We'll extend the notion of inverse functions in a later chapter to accommodate a possible mismatch of domain and range.

Exercise. Suppose $f : X \to X$ has an inverse function. Explain why $f(f^{-1}(x)) = x$ and why $f^{-1}(f(x)) = x$.

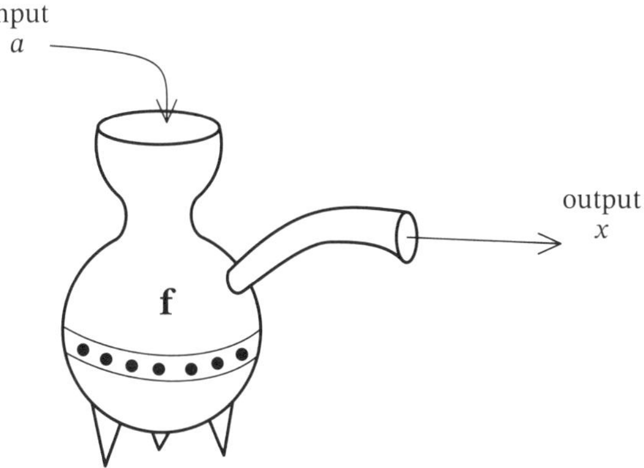

FIGURE 5

Solution. By definition, $f^{-1}(x)$ is the element that gives x as an output. So if we apply f to $f^{-1}(x)$, we get the output x! That is, $f(f^{-1}(x)) = x$. By definition, $f^{-1}(f(x))$ is the input that gives the output $f(x)$. Clearly x has output $f(x)$, and so $f^{-1}(f(x))$ is x. (In short, running a function machine backwards and then forwards or forwards and then backwards should land you back where you started.) $\quad\square$

Assuming we can run a value x backwards through a machine twice, it is reasonable to denote the final result as $f^{-2}(x)$. This is the composite function $f^{-1} \circ f^{-1}$ applied to x. In general, f^{-k} is the function f^{-1} composed with itself k times.

One can reason that $f^a \circ f^b = f^{a+b}$ holds even if a and b are integers.

Exercise. Can one give meaning to $f^{\frac{1}{2}}$? Presumably, this is a function with the property that $f^{\frac{1}{2}} \circ f^{\frac{1}{2}} = f$.

Solution. This is a surprisingly subtle question. See the essay "Compositional Square Roots"[1] for some thoughts on this. $\quad\square$

Comment. Having just learned that the standard notation for iteration is abandoned when it comes to trigonometric functions ($\cos^3(x)$, for instance, means the value $\cos(x)$ cubed), the notation for inverse functions

---

[1] www.jamestanton.com/wp-content/uploads/2012/03/Cool-Math-Essay_MAY-2015_Compositional-Square-Roots_v2.pdf

is *not* abandoned: $\sin^{-1}(x)$ does represent the inverse operation to the sine function, finding an input (angle) whose output is x.

To summarize: If k is a positive integer, then

$\sin^{k}(x)$ follows the algebra interpretation (it means $\sin(x)$ raised to the kth power) and not the iteration interpretation,

but

$\sin^{-1}(x)$ follows the iteration interpretation (it means the inverse to sine) and not the algebra interpretation (which would be $\frac{1}{\sin(x)}$).

Very confusing!

By the way, I personally do not know what $\sin^{-2}(x)$ would mean.

Multi-Valued Functions

Consider the operation *Sqrt* that assigns to each square number 1, 4, 9, 16, … its square roots. So to the number 9 we assign the set $\{-3, 3\}$.

Some might argue that this operation is not a function since for each input we are assigning more than one output. But we can view this operation as a function. We have

$$\text{Sqrt} : \text{Square Numbers} \rightarrow \text{Sets of Integers}$$

with the rule: assign to each square number its set of square roots.

A multi-valued "function" is a function: it's just one whose outputs are sets.

Some textbooks make a distinction between a function and a *relation*, with any operation that might assign more than one output to a given input being called a relation. But even these relations can be viewed as functions and there is no need for fussing.

Comment. For those who like formal thinking: a *function* from a set X to a set Y is a subset $f \subseteq X \times Y$ with the property that each x in X appears as a first element of precisely one pair in f. A *relation* from a set X to a set Y is subset $R \subseteq X \times Y$ with the property that each x in X appears as a first element of at least one pair in R.

MAA Problems

In each of these featured problem sections I give an account of my personal path to solving the given problem, sharing with you my human reactions and thoughts along the way. You, no doubt, will have a different set of reactions to each of these challenges and will develop alternative ways to solve them. Thus, you will have your own human mathematical experience!

MAA Featured Problem

> **(#22, AMC 12A, 2007)**
>
> For each positive integer n, let $S(n)$ denote the sum of the digits of n. For how many values of n is $n + S(n) + S(S(n)) = 2007$?
>
> > (A) 1
> > (B) 2
> > (C) 3
> > (D) 4
> > (E) 5

A Personal Account of Solving This Problem

Curriculum Inspirations Strategies (`www.maa.org/ci`):

> Strategy 2: Do something!

> Strategy 7: Perseverance is key.

I need to get a feel for the question. It seems confusing.

Let me try $n = 207$ (I don't know why I chose that number) and work out $n + S(n) + S(S(n))$ for it.

$$207 + S(207) + S(S(207)) = 207 + 9 + S(9) = 207 + 9 + 9 = 225.$$

Okay, this is not 2007. We're looking for numbers that give 2007.

Could a three-digit number work? Could $n = 100a + 10b + c$? This would give

$$n + S(n) + S(S(n)) = 100a + 10b + c + a + b + c + S(a + b + c),$$

and this would never equal a number in the 2000's. (Each of a, b, and c is at most 9.)

All right. So n needs to be a four-digit number: $n = 1000a + 100b + 10c + d$. (I can see that five digits would be too many.) I also see that $a = 1$ or 2.

So what can I say?

$$n = 1000a + 100b + 10c + d, \text{ with } a = 1 \text{ or } 2.$$

$$S(n) = a + b + c + d, \text{ and this is at most } 2 + 9 + 9 + 9 = 29.$$

$S(S(n)) = S(a + b + c + d)$. Of all the values from 1 to 29 the quantity $a + b + c + d$ might take, we see that 29 has the largest digit sum. So $S(S(n)) \leq 11$.

We also have

$$n + S(n) + S(S(n)) = 2007.$$

We want the possible values of n.

Oh! We have

$$n = 2007 - S(n) - S(S(n));$$

it is 2007 take away some values, and so $n \leq 2006$. Also, since $S(n) \leq 29$ and $S(S(n)) \leq 11$, it follows that $n \geq 2007 - 40 = 1967$. So n is one of the values 1967, 1968, $\ldots$, 2006.

But this is a lot of values to check!

Well, I can see pretty quickly that among $n = 2000, 2001, \ldots, 2006$ work, only $n = 2001$ works. So that is one possible value of n. Some success!

Let's assume now that we're dealing with $a = 1$ and $b = 9$.

Hmm.

Well, I can see that no number in the 1990's will work.

A number in the 1980's?

If $a = 1, b = 9$, and $c = 8$, then

$$n + S(n) + S(S(n)) = (1980 + d) + (18 + d) + S(18 + 9)$$
$$= 1998 + 2d + S(18 + d).$$

For this to equal 2007 we need $S(18 + d) = 9 - 2d$. I see that $d = 0$ and $d = 3$ work. So we have two more possible values for n, namely, 1980 and 1983.

In the 1970's?

If $a = 1, b = 9$, and $c = 7$, then

$$n + S(n) + S(S(n)) = (1970 + d) + (17 + d) + S(17 + 9)$$
$$= 1987 + 2d + S(17 + d),$$

meaning we need

$$S(17 + 2d) = 20 - 2d.$$

Only $d = 7$ works, giving $n = 1977$.

In the 1960's?

We only need to check 1967, 1968, and 1969 and none of these work.

So there are four possible values of n and the answer is (D).

Additional Problems

1. (#2, AMC 12, 2001) Let $P(n)$ and $S(n)$ denote the product and the sum, respectively, of the digits of the integer n. For example, $P(23) = 6$ and $S(23) = 5$. Suppose N is a two-digit number such that $N = P(N) + S(N)$. What is the units digit of N?

   (A) 2

   (B) 3

   (C) 6

   (D) 8

   (E) 9

2. (#13, AMC 10B, 2003) Let $\clubsuit(x)$ denote the sum of the digits of the positive integer x. For example, $\clubsuit(8) = 8$ and $\clubsuit(123) = 1+2+3 = 6$. For how many two-digit values x is $\clubsuit(\clubsuit(x)) = 3$?

   (A) 3

   (B) 4

   (C) 6

   (D) 9

   (E) 10

2

Sequences as Functions on $\mathbb{N}$

A sequence is a list of numbers, with a first number in the list, a second number in the list, and so on.

Example. The list of numbers 1, 4, 9, 16, ..., with an implied pattern, is a sequence. The first number in the list is 1, the fourth is 16, and, if the pattern is to be believed, the 19th number is 361.

The individual numbers in a sequence are called the *terms* of the sequence.

We can view a sequence as a function from the set of positive integers $\{1, 2, 3, \dots\}$ to the set of real numbers as given by the rule: *assign to the number n the nth number in the list.*

Example. In the above example, it seems we have the function that assigns to the input n, a positive whole number, the output n^2

Comment. Patterns need not be believed. For example, the next number in the sequence 1, 4, 9, 16, ... could well be 17. Maybe these numbers are following the formula

$$(-n^4 + 10n^3 - 32n^2 + 50n - 24)/3$$

instead.

This idea of thinking of a list as a function is simple enough. But matters are a bit confusing when it comes to developing a precise notation for this interplay. For example, for the list of square numbers, we could begin the list 0, 1, 4, 9, 16, ..., for which it would be handy to regard the first term in the list not as the first term but as the "zeroth term". Also, the rule "assign to n the number n^2" is valid for negative integers too, so we could talk of the (-3)rd square number, it is 9, and the (-19)th square number, which is 361. Thinking this way then gives us a "double ended" list of numbers, ..., 9, 4, 1, 0, 1, 4, 9, ..., which we still might want to regard as a sequence.

So it is possible to view a sequence as a function with domain $\{1, 2, 3, ...\}$ or with domain $\{0, 1, 2, 3, ...\}$ or possibly with domain $\{..., -3, -2, -1, 0, 1, 2, 3, ...\}$.

Comment. The set $\{..., -3, -2, -1, 0, 1, 2, 3, ...\}$ is called the set of *integers* and is denoted $\mathbb{Z}$ (from the German word *Zahlen* for numbers). There is some ambiguity as to what to call the sets $\{1, 2, 3, ...\}$ and $\{0, 1, 2, 3, ...\}$. Some school textbook authors call $\{1, 2, 3, ...\}$ the set of *natural numbers* and denote it $\mathbb{N}$ and they call $\{0, 1, 2, 3, ...\}$ the set of whole numbers (with no special symbol for this set). However, many mathematicians consider zero a natural number and so regard $\mathbb{N}$ to be the set $\{0, 1, 2, 3, ...\}$ and like to call negative integers whole numbers too. (In their minds, there is no difference between the set of whole numbers and the set of integers.)

Whether or not zero is regarded as a natural number and whether or not a negative integer is deemed a whole number depends on the author of the mathematical piece you are reading. There is no uniform agreement on this.

The set of all positive and negative integers and all positive and negative finite and infinite decimals is the set of real numbers, denoted $\mathbb{R}$. Everyone agrees on the meaning of this set.

So, in summary, a sequence is a function from $\mathbb{N}$ to $\mathbb{R}$ either with or without 0 regarded as an element of $\mathbb{N}$, or a function from $\mathbb{Z}$ to $\mathbb{R}$.

In this setting, if a sequence is abstractly given by the list a_1, a_2, a_3, then we can regard it as the function $a : \mathbb{N} \to \mathbb{R}$ which assigns to n the nth number a_n.

Comment. This sentence can be compactly written "$a(n) = a_n$", but writing that is probably more confusing than enlightening.

Most people do not think of sequences as functions and will simply write $a_1, a_2, a_3, ...$ or $\{a_n\}$ or simply $\{a\}$ or perhaps (a_n) or (a) or sometimes just a_n for a list of numbers with the letter a as the general name of the sequence. Whether the list starts with a first term or a zeroth term or applies to negative integer terms is to be decided by the context of the discussion.

Example. $\{n^2\}_{n=-3}^{\infty}$ is fancy notation for the sequence of square numbers starting with the square number $(-3)^2 = 9$.

Often the terms of a sequence are given by an explicit formula. For example, someone might write the following:

Consider the sequence $\{b_n\}$ given by $b_n = \dfrac{n}{n^2+1}$.

Here we can see, for example, that the 20th term of this sequence is $\dfrac{20}{401}$. (It is not clear from this statement whether this sequence starts with $n = 1$ or $n = 0$ or whether negative integer inputs are being allowed.)

Alternatively, an author might describe a sequence as given by a particular pattern.

A sequence a_n begins 1, 4, 13, 40, … for $n = 1, 2, 3, \ldots$ with each term one more than triple the previous term.

In this example we have

$$a_1 = 1,$$
$$a_2 = 3 \times 1 + 1 = 3 + 1,$$
$$a_3 = 3(3 + 1) + 1 = 3^2 + 3 + 1,$$
$$a_4 = 3(3^2 + 3 + 1) + 1 = 3^3 + 3^2 + 3 + 1,$$

suggesting that $a_n = 3^{n-1} + 3^{n-2} + \cdots + 3 + 1 = \dfrac{3^n - 1}{3 - 1} = \dfrac{1}{2}(3^n - 1)$. This is an explicit formula for the terms of the sequence. (We explain how to compute a sum of powers at the end of this chapter.)

A sequence defined by a description of how each term in the sequence is constructed from earlier terms is said to be defined *recursively*. It is not usually obvious—nor always possible—to construct an explicit formula for the terms of a sequence defined recursively.

Example. The Fibonacci numbers are given recursively as follows. Set

$$F_0 = 0,$$
$$F_1 = 1,$$

and

$$F_n = F_{n-1} + F_{n-2} \quad \text{for } n \geq 2.$$

Here we are being told that the sequence $F_0, F_1, F_2, F_3, \ldots$ begins 0, 1, … with each term beyond the first pair the sum of the two terms just before it.

It is not at all obvious if there might be an explicit formula for the Fibonacci numbers. (There is one! See `www.jamestanton.com/?p=602`.)

Additive and Multiplicative Structures, and Averages

Consider a rectangle with sides of length a units and b units (see Figure 6). Two questions:

> What is the side length of a square with the same perimeter as this rectangle?

> What is the side length of a square with the same area as this rectangle?

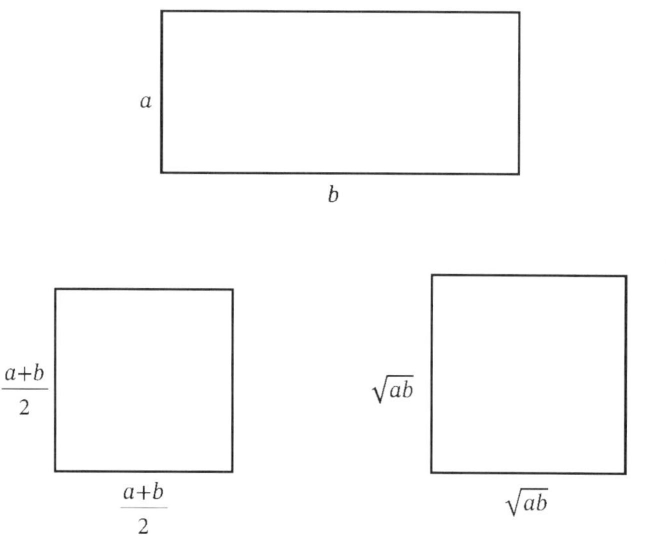

$$FIGURE\ 6$$

To match the perimeter of the rectangle, the square must have side length $\frac{a+b}{2}$. This is the *arithmetic mean* of a and b. To match the area of the rectangle, the square must have side length $\sqrt{ab}$, which is called the *geometric mean* of a and b.

Each mean is an "average" value of the two original values a and b: the first in some additive sense and the second in some multiplicative sense.

Comment. We can extend the notions of arithmetic and geometric means to more than two numbers. For a set on n values $a_1, a_2, \ldots, a_n$, we define their arithmetic mean to be

$$\frac{a_1 + a_2 + \cdots + a_n}{n}$$

and their geometric mean to be

$$\sqrt[n]{a_1 \cdot a_2 \cdot \cdots \cdot a_n}$$

(assuming, here, that each of the values is positive).

If a, x, and b are three numbers with x some mean of a and b, then the "difference" d between a and x is the same "difference" d between x and b.

Thinking additively with x the arithmetic mean, this means that $x = a + d$ and $b = x + d$ showing that our three numbers are $a, a + d$, and $a + 2d$ (and $a + d$ is indeed the arithmetic mean of a and $a + 2d$). The three numbers change by a constant additive factor d.

Thinking instead multiplicatively, with x the geometric mean, this means that $x = a \times d$ and $b = x \times d$ showing that our three numbers are a, ad, and ad^2 (and ad is indeed the geometric mean of a and ad^2). The three numbers change by a constant multiplicative factor d.

It is often helpful to keep the parallels between additive and multiplicative thinking close to mind.

Sequences with an Additive Structure

A sequence is said to be *arithmetic* if each term in the sequence beyond the first is the arithmetic mean of its two neighbors. This means that any three consecutive terms differ by the same additive factor d and so the sequence has the form

$$a, a + d, a + 2d, a + 3d, a + 4d, \ldots$$

for some numbers a and d. The value d is the difference between any two consecutive terms.

If a is regarded as the first term of the sequence, then the nth term of the sequence is given by $a + (n-1)d$. If, on the other hand, a is regarded as the zeroth term, then the nth term of the sequence is given by $a + nd$.

We also have the recursive definition of an arithmetic progression:

$$a_{n+1} = a_n + d \quad \text{with } a_1 = a \text{ (or perhaps } a_0 = a).$$

Comment. Admittedly I led us to the definition of an arithmetic sequence (also called an *arithmetic progression*) in a contorted way. A straightforward definition would be to call a sequence *arithmetic* if all consecutive terms in the sequence differ by the same constant value. But it is enlightening to realize that this is equivalent to saying that each term, beyond the first, is the arithmetic mean of its two neighbors.

Exercise. Show that each term in an arithmetic sequence is the arithmetic mean of its six immediate neighbors: three to the left and three to the right.

Solution. If x is a term in an arithmetic sequence with constant difference d, then its left three neighbors are $x - 3d, x - 2d, x - d$ and its right three neighbors are $x + d$, $x + 2d$, and $x + 3d$. These six values have arithmetic mean x. $\qquad\square$

Exercise. The number 175 is the sum of seven consecutive odd numbers. What is the first of those odd numbers?

Solution. The sequence of odd numbers is an arithmetic progression with constant difference 2. Being a bit clever with symmetry, we have

$$(x - 6) + (x - 4) + (x - 2) + x + (x + 2) + (x + 4) + (x + 6) = 175$$

for some integer x. We see that $7x = 175$ and so $x = 25$. The smallest odd number in this sum is 19. $\qquad\square$

Arithmetic sequences are used to model a population whose size changes over regular units of time by a fixed amount. Here is an unrealistic example.

Example. A town of a population 1050 declines by a count of 75 each month. Suppose we want to know in how many months the population of the town will be 300.

The population of the town after n months is given by $1050 - 75n$. And this equals 300 when $75n = 750$, that is, when $n = 10$.

$$n+1$$

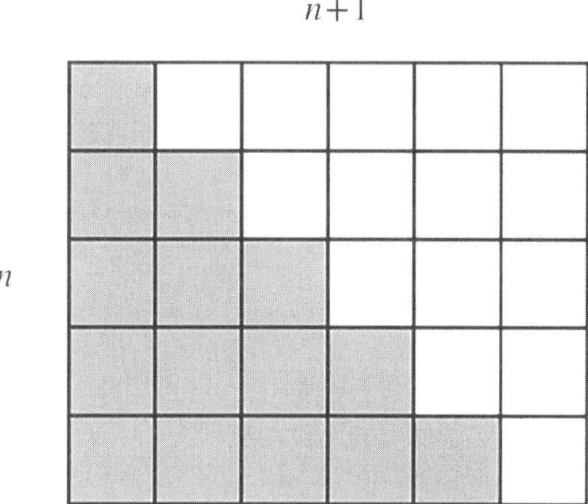

FIGURE 7

The counting numbers form a simple arithmetic progression. Figure 7 shows that the sum of the first n counting numbers $1 + 2 + \cdots + n$ corresponds to half the area of an $n \times (n + 1)$ rectangle. Thus we have

$$1 + 2 + 3 + \cdots + n = \frac{1}{2}n(n + 1).$$

One can compute the sum of terms of any finite set of terms in arithmetic progression by keeping this formula (better yet, image) in mind. For example, the sum of all three-digit multiples of seven can be computed as follows.

$$\begin{aligned}
105 + 112 + 119 + \cdots + 994 \\
= (98 + 1 \cdot 7) + (98 + 2 \cdot 7) + \cdots + (98 + 128 \cdot 7) \\
= 128 \cdot 98 + 7(1 + 2 + \cdots + 128) \\
= 12544 + 7 \times \frac{1}{2} \cdot 128 \cdot 129 \\
= 70336.
\end{aligned}$$

Comment. A sum of consecutive terms of a sequence is called a *series.*

Sequences with a Multiplicative Structure

A sequence is said to be *geometric* if each term in the sequence beyond the first is the geometric mean of its two neighbors. This means that any three consecutive terms differ by the same multiplicative factor d and so the sequence has the form

$$a, ad, ad^2, ad^3, ad^4, \ldots$$

for some numbers a and d. The value d is the ratio between any two consecutive terms.

If a is regarded as the first term of the sequence, then the nth term of the sequence is given by ad^{n-1}. If, on the other hand, a is regarded as the zeroth term, then the nth term of the sequence is given by ad^n.

We also have the recursive definition of an geometric sequence:

$$a_{n+1} = a_n \cdot d \quad \text{with } a_1 = a \text{ (or perhaps } a_0 = a\text{)}.$$

Comment. A geometric sequence (also called a *geometric progression*) is a sequence whose consecutive terms have the same ratio. This direct definition is equivalent to saying that each term, beyond the first, is the geometric mean of its two neighbors. Most texts prefer to use the symbol r (for "ratio") instead of d (for "difference") when writing an abstract geometric sequence and so write $a, ar, ar^2, ar^3, \ldots$.

Exercise. Show that each term in a geometric sequence with positive terms is the geometric mean of its six immediate neighbors: three to the left and three to the right.

Solution. The geometric mean of k numbers is the kth root of their product. If x is a term in a geometric sequence with constant nonzero ratio r, then its left three neighbors are $\frac{x}{r^3}, \frac{x}{r^2}, \frac{x}{r}$ and its right three neighbors are xr, xr^2, and xr^3. These six values have a product x^6 with sixth root x, as claimed. $\qquad\square$

A geometric progression is used to model a population or a count whose size changes by a fixed factor over regular units of time. Here are two unrealistic examples.

Example. A bacteria mass in a Petri dish currently weights 360 micrograms and declines in mass 2% every day. After n days, its mass can be approximated as

$$360(1 - 0.02)^n = 360 \times 0.98^n \text{ micrograms.}$$

Example. A savings account offers 5% annual interest calculated and applied monthly. With an initial balance of $1000, the balance of the account after m months is

$$1000 \times \left(1 + \frac{0.05}{12}\right)^m \text{ dollars.}$$

In algebra one learns that, for each positive integer n, the polynomial $x^n - 1$ is divisible by $x - 1$. We have

$$\frac{x^n - 1}{x - 1} = 1 + x + x^2 + \cdots + x^{n-1}.$$

(See Lesson 6.2 onwards of `www.gdaymath.com/courses/exploding-dots/`, for instance.) This allows us to compute the sum of the first n terms of a geometric progression:

$$a + ar + ar^2 + \cdots + ar^{n-1} = a(1 + r + r^2 + \cdots + r^{n-1}) = a\left(\frac{r^n - 1}{r - 1}\right).$$

The Geometric Series Formula

Consider the following sharing activity.

Suppose that you have a cup of juice to share between you and four friends. Dividing up the liquid equally gives one-fifth of a cup of liquid per person.

You next share your portion equally among the five of you. This leaves you with a fifth of a fifth, that is, $\frac{1}{25}$ of a cup of liquid, but each friend now has $\frac{1}{5} + \frac{1}{25}$ of a cup of juice.

You share your portion again. This leaves you with $\frac{1}{125}$ of a cup of liquid and each friend has $\frac{1}{5} + \frac{1}{25} + \frac{1}{125}$ of a cup of juice.

You keep doing this, indefinitely, beyond the end of time. The amount of juice you possess dwindles to zero, and so all the juice is equally distributed among your four friends. They must each possess

$\frac{1}{4}$ of a cup of juice. But the amount of liquid each person received is given by the (infinite) sum $\frac{1}{5} + \frac{1}{25} + \frac{1}{125} + \frac{1}{625} + \cdots$, the sum of the powers of a fifth. We conclude then that

$$\frac{1}{5} + \left(\frac{1}{5}\right)^2 + \left(\frac{1}{5}\right)^3 + \cdots = \frac{1}{4}.$$

In general, we can argue that

$$\frac{1}{n} + \left(\frac{1}{n}\right)^2 + \left(\frac{1}{n}\right)^3 + \cdots = \frac{1}{n-1}$$

holds for any integer n greater than 1. (Share a cup of liquid with $n-1$ friends.)

One proves in a calculus class that

$$1 + x + x^2 + x^3 + \cdots = \frac{1}{1-x}$$

is a valid formula for any value x whose size is less than 1 and our sharing argument gives the formula for $x = \frac{1}{n}$.

Comment. The formula $1 + x + x^2 + x^3 + \cdots = \frac{1}{1-x}$ is called the *geometric series formula* and it can also be established purely algebraic by dividing the expression 1 by the expression $1 - x$. (See Lesson 7.2 of `www.gdaymath.com/courses/exploding-dots/` for details.) Also, if you believe that the expression $1 + x + x^2 + x^3 + \cdots$ should have a meaningful value, call that value S, say, then pure algebra tells us what that value must be:

$$S = 1 + x + x^2 + x^3 + \cdots$$
$$= 1 + x(1 + x + x^2 + \cdots)$$
$$= 1 + xS.$$

This gives $S - xS = 1$, forcing us to conclude that $S = \frac{1}{1-x}$. (The work of calculus is to show that the infinite sum does have a meaningful answer for certain values of x.)

MAA Featured Problem

(#24, AMC 10B, 2003)

The first four terms of an arithmetic sequence are $x + y$, $x - y$, xy, and x/y, in that order. What is the fifth term?

(A) $-\dfrac{15}{8}$

(B) $-\dfrac{6}{5}$

(C) 0

(D) $\dfrac{27}{20}$

(E) $\dfrac{123}{40}$

A Personal Account of Solving This Problem

Curriculum Inspirations Strategy (`www.maa.org/ci`):

> Strategy 2: Do something!

As I read this question, my eyes skip over the algebraic expressions. But I do notice we are talking about numbers in an arithmetic sequence. And I recall that a sequence of numbers is arithmetic if the terms change by a constant amount.

What are our numbers in the sequence? They are $x + y$, $x - y$, xy, and x/y.

To go from $x + y$ to $x - y$ we subtract $2y$. So to go from $x - y$ to xy we must also subtract $2y$. (The difference must be constant.) This gives $x - y - 2y = xy$. That is, $x - 3y = xy$.

To go from xy to x/y we must again subtract $2y$, so $xy - 2y = \dfrac{x}{y}$. Let's rewrite this as $xy^2 - 2y^2 = x$.

So we have the two equations

$$x - 3y = xy,$$
$$xy^2 - 2y^2 = x.$$

I suppose we can try to solve for x and y. But is that what the question wants?

What is the fifth term?

Okay. We want the value of the next term of the sequence, which is $\frac{x}{y} - 2y$. If we know the values of x and y, then we know the value of this fifth term. I guess we do have to find x and y.

The two equations we have look scary. Let's work with the first one since it involves no squared terms.

From $x = xy + 3y$ we get $x = y(x + 3)$ and so $y = \frac{x}{x+3}$. We could substitute this into the second equation, but that would be horrid!

Okay. What if I solved for x instead? From $x - xy = 3y$ we get $x(1 - y) = 3y$ and so $x = \frac{3y}{1-y}$. Putting this into the second equation produces

$$\frac{3y^3}{1 - y} - 2y^2 = \frac{3y}{1 - y}.$$

Multiplying through by $1 - y$ yields

$$3y^3 - 2y^2 + 2y^3 = 3y.$$

That is,

$$5y^3 - 2y^2 - 3y = 0.$$

Dividing through by y gives $5y^2 - 2y - 3 = 0$. (Ooh! What if $y = 0$? That's a possible solution!)

Looking at $5y^2 - 2y - 3 = 0$, we see that $y = 1$ works. This means we can write

$$5y^2 - 2y - 3 = 0,$$
$$(y - 1)(\text{something}) = 0,$$
$$(y - 1)(5y + 3) = 0.$$

So $y = -\frac{3}{5}$ is another possible solution. We have $y = 0$ or $y = 1$ or $y = -3/5$.

Since $x = \frac{3y}{1-y}$, we can't have $y = 1$. So either $y = 0$ and $x = 0$ or $y = -3/5$ and $x = \frac{(-9/5)}{1+3/5} = \frac{-9}{5+3} = -\frac{9}{8}$. The fifth term is $\frac{x}{y} - 2y$, so $y \neq 0$.

The fifth term must thus be

$$\frac{(-9/8)}{(-3/5)} - 2\left(-\frac{3}{5}\right) = \frac{45}{24} + \frac{6}{5} = \frac{369}{120} = \frac{123}{40}.$$

The answer is (E).

Additional Problems

3. (#6, AMC 12B, 2003) The second and fourth terms of a geometric sequence are 2 and 6. Which of the following is a possible first term?

   (A) $-\sqrt{3}$

   (B) $-\dfrac{2\sqrt{3}}{3}$

   (C) $-\dfrac{\sqrt{3}}{3}$

   (D) $\sqrt{3}$

   (E) 3

4. (#7, AMC 12A, 2007) Let a, b, c, d, and e be five consecutive terms in an arithmetic sequence, and suppose that $a + b + c + d + e = 30$. Which of the following can be found?

   (A) a

   (B) b

   (C) c

   (D) d

   (E) e

5. (#19, AMC 10B, 2002) Suppose that $\{a_n\}$ is an arithmetic sequence with $a_1 + a_2 + \cdots + a_{100} = 100$ and $a_{101} + a_{102} + \cdots + a_{200} = 200$. What is the value of $a_2 - a_1$?

   (A) 0.0001

   (B) 0.001

   (C) 0.01

   (D) 0.1

   (E) 1

6. (#23, AMC 10B, 2002) Let $\{a_k\}$ be a sequence of integers such that $a_1 = 1$ and $a_{m+n} = a_m + a_n + mn$, for all positive integers m and n. What is a_{12}?

   (A) 45

   (B) 56

   (C) 67

   (D) 78

   (E) 89

7. (#24, AMC 10A, 2004) Let $a_1, a_2, \ldots$ be a sequence with the following properties:

(i) $a_1 = 1$, and

(ii) $a_{2n} = n \cdot a_n$ for any positive integer n.

What is the value of $a_{2^{100}}$?

(A) 1

(B) 2^{99}

(C) 2^{100}

(D) 2^{4950}

(E) 2^{9999}

8. (#19, AMC 10B, 2004) In the sequence 2001, 2002, 2003, ..., each term after the third is found by subtracting the previous term from the sum of the two terms that precede that term. For example, the fourth term is $2001 + 2002 - 2003 = 2000$. What is the 2004th term in this sequence?

(A) -2004

(B) -2

(C) 0

(D) 4003

(E) 6007

9. (#11, AMC 10B, 2005) The first term of a sequence is 2005. Each succeeding term is the sum of the cubes of the digits of the previous term. What is the 2005th term of the sequence?

(A) 29

(B) 55

(C) 85

(D) 133

(E) 250

10. (#18, AMC 10B, 2006) Let $a_1, a_2, \ldots$ be a sequence for which $a_1 = 2$, $a_2 = 3$, and $a_n = \dfrac{a_{n-1}}{a_{n-2}}$ for each positive integer $n \geq 3$. What is a_{2006}?

(A) $\dfrac{1}{2}$

(B) $\dfrac{2}{3}$

(C) $\dfrac{3}{2}$

(D) 2

(E) 3

11. (#10, AMC 12A, 2004) The sum of 49 consecutive integers is 7^5. What is their median?

(A) 7

(B) 7^2

(C) 7^3

(D) 7^4

(E) 7^5

12. (#9, AMC 12B, 2002) Let a, b, c, and d be positive real numbers such that a, b, c, d form an increasing arithmetic sequence and a, b, d form a geometric sequence. What is a/d?

(A) $\dfrac{1}{12}$

(B) $\dfrac{1}{6}$

(C) $\dfrac{1}{4}$

(D) $\dfrac{1}{3}$

(E) $\dfrac{1}{2}$

13. (#15, AMC 12B, 2007) The geometric series $a + ar + ar^2 + \cdots$ has a sum of 7, and the terms involving odd powers of r have a sum of 3. What is $a + r$?

(A) $\dfrac{4}{3}$

(B) $\dfrac{12}{7}$

(C) $\dfrac{3}{2}$

(D) $\dfrac{7}{3}$

(E) $\dfrac{5}{2}$

14. (#14, AMC 12A, 2004) A sequence of three real numbers forms an arithmetic progression with first term 9. If 2 is added to the second term and 20 is added to the third term, the three resulting numbers form a geometric progression. What is the smallest possible value for the third term of the geometric progression?

(A) 1

(B) 4

(C) 36

(D) 49

(E) 81

15. (#21, AMC 12A, 2002) Consider the sequence of numbers 4, 7, 1, 8, 9, 7, 6, …. For $n > 2$, the nth term of the sequence is the units digit of the sum of the two previous terms. Let S_n denote the sum of the first n terms of this sequence. What is the smallest value of n for which $S_n > 10{,}000$?

(A) 1992

(B) 1999

(C) 2001

(D) 2002

(E) 2004

16. (#21, AMC 12B, 2002) For all positive integers n less than 2002, let

$$a_n = \begin{cases} 11, & \text{if } n \text{ is divisible by 13 and 14,} \\ 13, & \text{if } n \text{ is divisible by 14 and 11,} \\ 14, & \text{if } n \text{ is divisible by 11 and 13,} \\ 0, & \text{otherwise.} \end{cases}$$

What is $\sum_{n=1}^{2001} a_n$?

(A) 448

(B) 486

(C) 1560

(D) 2001

(E) 2002

Comment. The Greek letter $\sum$ for "S" is used in mathematics to denote a sum. Here $\sum_{n=1}^{2001} a_n$ means the sum $a_1 + a_2 + \cdots + a_{2001}$.

17. (#25, AMC 12, 2001) Consider sequences of positive real numbers of the form $x, 2000, y, \ldots$ in which every term after the first is 1 less than the product of its two immediate neighbors. For how many different values of x does the term 2001 appear somewhere in the sequence?

(A) 1

(B) 2

(C) 3

(D) 4

(E) more than 4

18. (#25, AMC 12B, 2006) A sequence $a_1, a_2, \ldots$ of nonnegative integers is defined by the rule $a_{n+2} = |a_{n+1} - a_n|$ for $n \geq 1$. If $a_1 = 999$, $a_2 < 999$, and $a_{2006} = 1$, how many different values of a_2 are possible?

    (A) 165

    (B) 324

    (C) 495

    (D) 499

    (E) 660

19. (#22, AMC 12B, 2005) A sequence of complex numbers $z_0, z_1, z_2, \ldots$ is defined by the rule

$$z_{n+1} = \frac{iz_n}{\bar{z}_n}$$

where $\bar{z}_n$ is the complex conjugate of z_n and $i^2 = -1$. Suppose that $|z_0| = 1$ and $z_{2005} = 1$. How many possible values are there for z_0?

    (A) 1

    (B) 2

    (C) 4

    (D) 2005

    (E) 2^{2005}

3
Numerical Functions on $\mathbb{R}$

School mathematics places a strong focus on numerical functions with domain the set of real numbers (or just a subset of the reals) and range within the set of real numbers too:

$$f : \mathbb{R} \to \mathbb{R}.$$

Here it is almost universally agreed to denote a general input of such a function with the symbol x and a general output with the symbol y. In writing

$$y = f(x)$$

we are making a claim that y equals the output $f(x)$ for a given input x. The claim might or might not be true.

Example. Suppose f is the squaring function: it assigns to each real number x its square x^2. Then $25 = f(5)$ is a true statement and "$-3 = f(x)$ for some x" is a false one.

It often happens that we can describe the rule of a numerical function by a formula and here people often muddle the meaning of an equation and the definition of a function.

An Aside on Equations

Mathematics is a language. For those reading this book, that language is English while reading it. For example, the statement "$2 + 4 = 6$" has a noun (the quantity two-plus-four), a verb (equals), and an object (six). The statement

$$\frac{1}{2} + \frac{1}{3} = \frac{3}{6} + \frac{2}{6} = \frac{5}{6}$$

is the sentence "One half plus one third is equivalent to three sixths plus two sixths, which is five sixths."

Some mathematical sentences are true ($2 + 4 = 6$, for instance) and some are false ($1 = 2$, for example). Mathematics is usually interested in sentences that are true.

But one usually needs a context to determine the truth value of a statement. For example, we cannot assess the truth of the claim "Suzy is less than five feet tall" without knowing which particular Suzy of the world we are referring to.

The same is true of the mathematical statements involving variables, such as "$x^2 = 25$". Some values of x make this a true statement; others do not. In and of itself, the statement $x^2 = 25$ has no truth value. (But we can't help but think of the numbers 5 and -5, the values that make this statement a true one about numbers. We are compelled to seek truth!)

In mathematics, an equation is any statement that claims that two expressions are equal. If variables are involved, we cannot say whether or not the equation speaks truth. The work of an algebra class is to find which values of the variable(s) involved make the given equation a true statement about numbers.

Defining Numerical Functions

One will often read in a textbook a statement such as

$$f(x) = 1 + \sqrt{x}.$$

This is an equation. It says, "The output of a function whose name is f is given by $1 + \sqrt{x}$ if x is the input to that function." Without knowing any details of the function f we have no way to assess whether or not this is a true or false statement, or which particular values of x make it true.

However, authors—including me—will write an equation intended not as an equation to be studied, but as a way of defining a function whose name is f.

Example. Without being told otherwise, the statement

$$h(x) = x^2 + x^3$$

is to be interpreted as: "A new function $h : \mathbb{R} \to \mathbb{R}$ is being considered. It takes an input x and assigns to it the sum of its square and cube, $x^2 + x^3$."

Comment. This abuse of mathematical notation can be confusing. Many mathematicians prefer to write

$$h : \mathbb{R} \to \mathbb{R},$$
$$x \mapsto x^2 + x^3$$

with an ordinary arrow to show between which two sets a function operates and a barred arrow to show how a generic input is transformed. This notational approach is much clearer, but it is rarely used in school texts.

The (potentially confusing) practice of writing an equation to imply the definition of a function goes deeper still. An equation such as $y = 2x + 3$, for instance, is just an equation. Some particular values for x and y make it a true statement about numbers; other pairs of values do not.

But because we have become so ingrained to think of x as an input and y as an output, many people, without thinking, will associate with this equation a function: the function from $\mathbb{R}$ to $\mathbb{R}$ that takes a generic input x and associates to it the output $2x + 3$. Notice that no name for this function is indicated.

Here is a summary of matters.

Even though

$$f(x) = [\text{some expression involving } x]$$

is an equation that might or might not be true depending on the value of x and what the function f does, such a statement is often interpreted as a definition of a function named f:

$$f : \mathbb{R} \to \mathbb{R},$$
$$x \mapsto \text{the value of that expression for that value of } x.$$

Even though

$$y = [\text{some expression involving } x]$$

is an equation that might or might not be true depending on the chosen values for x and y, such a statement is often interpreted as the definition of an unnamed function from $\mathbb{R}$ to $\mathbb{R}$ that takes an input x and assigns to it the value of that expression for that value of x.

Comment. The set of complex numbers is denoted $\mathbb{C}$. One can, of course, consider functions $f : \mathbb{C} \to \mathbb{C}$ too.

If a function is defined by a formula and there is no commentary about that formula, then it is assumed that the domain of the function is the subset of all real values for which that formula makes sense.

Example. If we are told simply that a function a is given by $a(x) = \dfrac{1}{x}$, then we should probably assume its domain is the set of all real numbers different from zero.

We can write

$$\text{Domain} = \{\, x \mid x \neq 0 \,\} \quad \text{or} \quad \{\, x \in \mathbb{R} \mid x \neq 0 \,\}.$$

This is read as "the set of all real numbers x such that x is different from zero".

(It is possible that we might be told later that we are to consider this function only on positive inputs, for example, in which case the domain of the function is just the set of all positive real numbers.)

Example. If $w(x) = \sqrt{x - 1}$, then we should assume that the domain of this function is the set of all real values greater than or equal to 1:

$$\text{Domain} = \{\, x \mid x \geq 1 \,\}.$$

Comment. There are multiple ways to express regions of the real number line. For example, the set of all nonnegative real numbers might be described simply that way, in words (there is no problem with writing words in mathematics!) or in set notation as $\{\, x \mid x \geq 0 \,\}$ or in interval notation $[0, \infty)$ or by shading a number line with a closed dot to represent the inclusion of that value (and an open dot to indicate the exclusion of that value):

Recall in interval notation that round brackets indicate "all values up to that end value, but not the end value itself", and square brackets indicate to include that end value.

Example. $[2, 3)$ is the set of all real numbers between 2 and 3, including 2 itself, but excluding 3.

$(-10, 5) = \{x \mid -10 < x < 5\}$.

$[8, 8]$ is just a single number.

$(8, 8]$ is the empty set.

An interval with left endpoint a and right endpoint b, as for (a, b) or $(a, b]$, for instance, is considered empty if $a > b$. For example, the set $[5, 4)$ is empty.

The symbols ∞ and $-\infty$ are used to indicate arbitrarily large values, in either a positive sense or a negative sense. Round brackets are always used with these symbols.

Example. $(-\infty, 2)$ is the set of all real numbers less than 2.

$[-1, \infty) = \{x \mid x \geq -1\}$.

$(-\infty, \infty) = \mathbb{R}$.

Despite all the possible notations for intervals of real numbers, it is always acceptable to just describe intervals in words.

Exercise. What is the largest possible domain of f given by $f(x) = 2 + \sqrt{x + 4}$, and for that domain, what is the range of the function?

Solution. The expression defining the function is meaningful only for $x \geq -4$. The largest possible domain is

$$
\begin{aligned}
\text{Domain} &= \text{set of all real numbers greater than or equal to } -4 \\
&= \{x \mid x \geq -4\} \\
&= [-4, \infty).
\end{aligned}
$$

(Any one of these descriptions of the domain is acceptable.)

What possible outputs can appear? Each output equals two plus a nonnegative quantity. (And we see that the input $x = -4$ gives the output 2 itself.) Every value 2 or greater does appear as an output.

$$
\begin{aligned}
\text{Range} &= \text{set of all real values greater than or equal to } 2 \\
&= \{y \mid y \geq 2\} \\
&= [2, \infty). \qquad \square
\end{aligned}
$$

Comment. It is common practice to use the symbol y when describing outputs. This is just a social convention. There is no mathematical reason to follow it.

MAA Problems

20. (#9, AMC 12B, 2007) A function f has the property that $f(3x-1) = x^2 + x + 1$ for all real numbers x. What is $f(5)$?

(A) 7

(B) 13

(C) 31

(D) 111

(E) 211

21. (#9, AMC 12, 2001) Let f be a function satisfying $f(xy) = f(x)/y$ for all positive real numbers x and y. If $f(500) = 3$, what is the value of $f(600)$?

(A) 1

(B) 2

(C) $\dfrac{5}{2}$

(D) 3

(E) $\dfrac{18}{5}$

22. (#16, AMC 12B, 2004) A function f is defined by $f(z) = i\bar{z}$, where $i = \sqrt{-1}$ and $\bar{z}$ is the complex conjugate of z. How many values of z satisfy both $|z| = 5$ and $f(z) = z$?

(A) 0

(B) 1

(C) 2

(D) 4

(E) 8

23. (#18, AMC 12A, 2006) The function f has the property that for each real number x in its domain, $1/x$ is also in its domain and $f(x) + f\left(\dfrac{1}{x}\right) = x$. What is the largest set of real numbers that can be in the domain of f?

(A) $\{x \mid x \neq 0\}$

(B) $\{x \mid x < 0\}$

(C) $\{x \mid x > 0\}$

(D) $\{x \mid x \neq -1 \text{ and } x \neq 0 \text{ and } x \neq 1\}$

(E) $\{-1, 1\}$

4

Composite Functions and Inverse Functions

Consider the function $f : \mathbb{R} \to \mathbb{R}$ given by $f(x) = x^2 + 1$. Then we have, for example,

$$f(3) = 10,$$
$$f(-3) = 10,$$
$$f(q) = q^2 + 1,$$
$$f(x + 1) = (x + 1)^2 + 1 = x^2 + 2x + 2,$$

and

$$f(f(x)) = (f(x))^2 + 1 = (x^2 + 1)^2 + 1 = x^4 + 2x^2 + 2.$$

The last three examples bring home the point that the name of the input is immaterial. It is best to read a defining formula such as $f(x) = x^2 + 1$ simply as instructions without reference to a specific input name. In our example we are told that the function f takes a numerical input, squares it, and adds 1 to create the matching output. This helps keep track of what to do when chosen inputs are themselves expressions (or, even more confusing, if the input is an output of the function).

Comment. We could have computed the final example as

$$f(f(x)) = f(x^2 + 1) = (x^2 + 1)^2 + 1.$$

Exercise. If $f(x) = x^2 + 1$ and $g(x) = |x| - 1$ each define functions from $\mathbb{R}$ to $\mathbb{R}$, determine the following expressions:

   (a) $g(f(2))$,

   (b) $f(g(2))$,

   (c) $g(f(x))$,

45

(d) $f(g(x))$,

(e) $g(g(\sqrt{f(x)-1}))$.

Solution.   (a) $g(f(2)) = g(5) = 4$ or $g(f(2)) = |f(2)|-1 = |5|-1 = 4$.

(b) $f(g(2)) = f(1) = 2$ or $f(g(2)) = (g(2))^2 + 1 = 1^2 + 1 = 2$.

(c) $g(f(x)) = g(x^2 + 1) = x^2$ or $g(f(x)) = f(x) - 1 = x^2$.

(d) $f(g(x)) = f(|x| - 1) = x^2 - 2|x| + 2$ or $f(g(x)) = (g(x))^2 + 1 = (|x| - 1)^2 + 1 = x^2 - 2|x| + 2$.

(e) $g(g(\sqrt{f(x)-1})) = g(g(\sqrt{x^2})) = g(g(|x|)) = g(|x| - 1)| = ||x| - 1| - 1$. (Is this final expression equivalent to $|x| - 2$?) $\square$

Given two numerical functions f and g and an input value x, the expression $g(f(x))$ matches the output of the composite function $g \circ f$. As we saw in the previous example there are two ways to think through the computation of $g(f(x))$. Sometimes one approach might be more manageable than the other.

Comment. Sometimes it can be convenient to regard a complicated expression defining a function as the result of a composition of functions. For example, consider

$$f(x) = \sqrt{(\sqrt{\sin(x) + 5} - 1)^2 + 7}.$$

We can see that if we set $u(x) = \sqrt{x}, v(x) = x^2 + 7, w(x) = \sqrt{x} - 1$, and $t(x) = \sin(x) + 5$, then we have $f(x) = u(v(w(t(x))))$.

Inverse Functions

A function is said to be *one-to-one* if no two inputs produce the same output.

Example. The function that S assigns to each American citizen a Social Security number should be one-to-one.

The function M that assigns to each person his or her biological mother is not one-to-one. (Two sisters, for instance, are assigned the same biological mother.)

If a function is one-to-one, then each output of the function arises from just one input.

Suppose a function $f : X \to Y$ is one-to-one. Then we can define a function f^{-1}, called the *inverse* function, that takes an element in the range of f and assigns to it the input from which it came:

$$f^{-1} : (\text{Range of } f) \to X,$$
$$y \mapsto \text{the input } x \text{ with } f(x) = y.$$

Example. The inverse function to the Social Security number function assigns to each Social Security number the American citizen with that number.

Suppose a function $f : X \to Y$ is one-to-one. Then any value y which is an output of the function has an input x from which it came. We have $f(x) = y$ (x is the input that gives y) and so $f^{-1}(y) = x$ (the input from which y came is x). The two statements $y = f(x)$ and $x = f^{-1}(y)$ are equivalent statements.

Example. If S is the Social Security number function, then

$$S(\text{Allister Fogglesnort the Third}) = 123\,456\,789$$

and

$$S^{-1}(123\,456\,789) = \text{Allister Fogglesnort the Third}$$

say the same thing. The logic, again, is tautological.

Mathematics classrooms are primarily concerned with numerical functions $f : \mathbb{R} \to \mathbb{R}$.

Example. Consider the numerical function given by $f(x) = 2x + 1$.

For the input $x = 1$, we obtain the output $y = 3$.

For the input $x = -7$, we obtain the output $y = -13$.

Pushing the function machine backwards

If $y = -5$ is an output, then $x = -3$ was the input.

If $y = 10$ is an output, then $x = 4\frac{1}{2}$ was the input.

Let's analyze this situation more generally. If y is an output of this function, then $y = 2x + 1$ for some input x. (That is, $y = f(x)$.) Some algebra shows that x must be given by $x = \frac{1}{2}(y - 1)$.

This clear and definite formula tells us three things.

(1) The range of f is all of $\mathbb{R}$. (For any real number y, the expression $\frac{1}{2}(y-1)$ is a meaningful number and it represents an input that gives the output y : $f\left(\frac{1}{2}(y-1)\right) = f(x) = y$.)

(2) The function f is one-to-one. (For any output y, there is only one possible input x from which it came, namely, $x = \frac{1}{2}(y-1)$.)

(3) The defining formula for the inverse function is $f^{-1}(y) = \frac{1}{2}(y-1)$.

Example. Consider $h : \{x \mid x \geq 0\} \to \{x \mid x \geq 0\}$ given by $h(x) = \sqrt{x+1} - 1$. If y is an output of this function, then $y = \sqrt{x+1} - 1$ for some input x. Algebra shows that $x = (y+1)^2 - 1$, a nonnegative real number. There is no ambiguity as to what this matching input to y must be. Thus the range of h is the set of all nonnegative reals, the function is one-to-one, and h^{-1} exists and is given by $h^{-1}(y) = (y+1)^2 - 1$.

Suppose $f : \mathbb{R} \to \mathbb{R}$ is a numerical function. If y is an output of the function, then $y = f(x)$ for some input x. If it is possible to algebraically manipulate the equation $y = f(x)$ to obtain an unambiguous, well-defined expression for x, then we are likely in a situation where we can argue that the inverse function f^{-1} exists and, moreover, have a defining formula for it.

Example. Consider the squaring function $s : \mathbb{R} \to \mathbb{R}$ given by $s(x) = x^2$.

For the input $x = 0.1$, we obtain the output $y = 0.01$.

For the input $x = \sqrt{\pi}$, we obtain the output $y = \pi$.

The reverse direction is problematic.

Given the output $y = 25$, say, there is no unique matching input. We could have $x = 5$ or -5.

This function is not one-to-one and so has no inverse.

However, this example can be salvaged if we restrict the domain of s. Let's regard s as a function from the set of nonzero real numbers to the set itself:

$$s : \{x \mid x \geq 0\} \to \{y \mid y \geq 0\},$$

$$x \mapsto x^2.$$

In this context, for the output $y = 25$ there is a unique input from which it came, namely, $x = 5$.

We see now that the inverse function to s now exists.

Going further: If y is an output of this restricted function, then $y = x^2$ for some nonnegative real number x. Consequently, $x = \sqrt{y}$, the nonnegative root of y. We have that $s^{-1}(y) = \sqrt{y}$.

Example. Consider the sine function. We have $\sin\left(\frac{\pi}{4}\right) = \frac{1}{\sqrt{2}}$, for instance. But this is not the only input that has this output:

$$\frac{1}{\sqrt{2}} = \sin\left(\frac{\pi}{4}\right) = \sin\left(\frac{3\pi}{4}\right) = \sin\left(\frac{9\pi}{4}\right) = \sin\left(\frac{11\pi}{4}\right) = \cdots.$$

If we restrict the sine function to the domain

$$\left\{ x \,\middle|\, -\frac{\pi}{2} \leq x \leq \frac{\pi}{2} \right\},$$

then an inverse function $\sin^{-1}$ exists. In this context,

$$\sin^{-1}\left(\frac{1}{\sqrt{2}}\right) = \frac{\pi}{4},$$

$$\sin^{-1}(0) = 0,$$

$$\sin^{-1}\left(-\frac{\sqrt{3}}{2}\right) = -\frac{\pi}{3},$$

and so on.

Exercise. A *linear function* $L : \mathbb{R} \to \mathbb{R}$ is given by $L(x) = ax + b$ for some constant values a and b with $a \neq 0$. Argue that linear functions are invertible.

Solution. If y is the output of a linear function, then $y = ax + b$ for some real number x. Then x is unambiguously given by $x = \frac{1}{a}(y - b)$. Thus each real number is an output of L and comes from just one input. The inverse function exists and is defined by $L^{-1}(y) = \frac{1}{a}(y - b)$. $\qquad\square$

Comment. In school texts, functions with straight-line graphs are called linear functions. Saying that a function is linear has a slightly different meaning in college mathematics.

We have seen that if $f(x) = 2x + 1$, then f is invertible and $f^{-1}(y) = \frac{1}{2}(y - 1)$. This formula takes an output value of f, called y, and shows how to convert it to its matching input. But y is an input to the function f^{-1} and society strongly prefers denoting input values with the letter x! So rather than write $f^{-1}(y) = \frac{1}{2}(y - 1)$, one might be obliged to write $f^{-1}(x) = \frac{1}{2}(x - 1)$. Just keep in mind that the symbol we choose to use for an input, or output, is immaterial: formulas that define functions should always be read as sets of general instructions not tied to any particular variable names.

Here is a complicated example.

Let f be the function defined on the set of all real numbers different from -1 by $f(x) = \frac{1}{1+x}$. Let g be the function defined on the set of nonzero reals by $g(x) = \frac{1}{x} - 1$. Then I claim that f and g are inverse functions. (That is, $f^{-1} = g$ and $g = f^{-1}$.) How could we see this?

First observe that zero is never an output value of f, but every nonzero real does appear as an output value. Also observe that -1 is never an output value of g, but every real value different from -1 is a possible output. We have

$$f : \mathbb{R}/\{-1\} \to \mathbb{R}/\{0\}$$

and

$$g : \mathbb{R}/\{0\} \to \mathbb{R}/\{-1\}.$$

Now look at $f(g(x))$ for x a nonzero real number. We have

$$f(g(x)) = \frac{1}{1 + g(x)} = \frac{1}{1 + \frac{1}{x} - 1} = x.$$

Now look at $g(f(x))$ for x different from -1. We have

$$g(f(x)) = \frac{1}{f(x)} - 1 = \frac{1}{\frac{1}{1+x}} - 1 = 1 + x - 1 = x.$$

This is curious. Both $f \circ g$ and $g \circ f$ take an input and leave it unchanged. This curiosity has some consequences.

(1) The function f must be one-to-one. (Could two different inputs x_1 and x_2 for f give the same output? Well, no. If $f(x_1) = f(x_2)$, then we'd have $g(f(x_1)) = g(f(x_2))$ as well. But this reads $x_1 = x_2$, showing that we didn't have two different inputs after all.)

(2) The function g must be one-to-one. (Could two different inputs x_1 and x_2 for g give the same output? Well, no. If $g(x_1) = g(x_2)$, then we'd have $f(g(x_1)) = f(g(x_2))$ as well. But this reads $x_1 = x_2$, showing that we didn't have two different inputs after all.)

(3) If y is an output of f, then $g(y)$ is the input from which it came. (We have $f(g(y)) = y$.) Thus g is acting as f^{-1}

(4) If y is an output of g, then $f(y)$ is the input from which it came. (We have $g(f(y)) = y$.) Thus f is acting as g^{-1}

Thus f and g are each invertible and they are each other's inverse function.

Comment. We could also identify the inverse function to f, say, by writing $y = \dfrac{1}{1+x}$ for an output value y and some matching input value x. Then algebra shows that $x = \dfrac{1}{y} - 1$. This is a clear and unambiguous value for x showing that f is one-to-one with inverse function defined by the same formula that defines g.

This complicated example shows that if X and Y are subsets of the real numbers and $f : X \to Y$ and $g : Y \to X$ are functions with the property that $g(f(x)) = x$ for each x in X and $f(g(x)) = x$ for each x in Y, then f and g are each invertible. Moreover, f is the inverse function to g and g is the inverse function to f.

Exercise. Let f be given by

$$f(x) = 1 + \frac{1}{x}.$$

Check that

$$f^2(x) = f(f(x))$$
$$= \frac{2x + 1}{x + 1},$$
$$f^3(x) = \frac{3x + 2}{2x + 1},$$
$$f^4(x) = \frac{5x + 3}{3x + 2}.$$

Find the values of $f(1)$, $f^2(1)$, $f^3(1)$, $f^4(1)$ and predict the value of $f^5(1)$. Compute $f^{-1}(x)$ and $f^{-2}(x)$. Does the function f have a fixed point (that is, a value x such that $f(x) = x$)?

Suggestion for a solution. Think Fibonacci and the golden ratio!

MAA Problems

24. (#9, AMC 12B, 2003) Let f be a linear function for which $f(6) - f(2) = 12$. What is $f(12) - f(2)$?

   (A) 12

   (B) 18

   (C) 24

   (D) 30

   (E) 36

25. (#13, AMC 12B, 2004) If $f(x) = ax + b$ and $f^{-1}(x) = bx + a$ with a and b real, what is the value of $a + b$?

   (A) -2

   (B) -1

   (C) 0

   (D) 1

   (E) 2

5 Graphing

Suppose we are wondering if taller people tend to have larger feet. To examine this question we might interview a number people and ask each person for his or her shoe size and height in inches. We can organize the data we obtain in a visual *scatter plot*. In the fictitious plot of Figure 8, each point plotted represents the two data values obtained from one individual: the horizontal coordinate is the individual's shoe size, and the vertical coordinate is her height. Visual examination of the plot will help us decide whether or not our original question is worth pursuing.

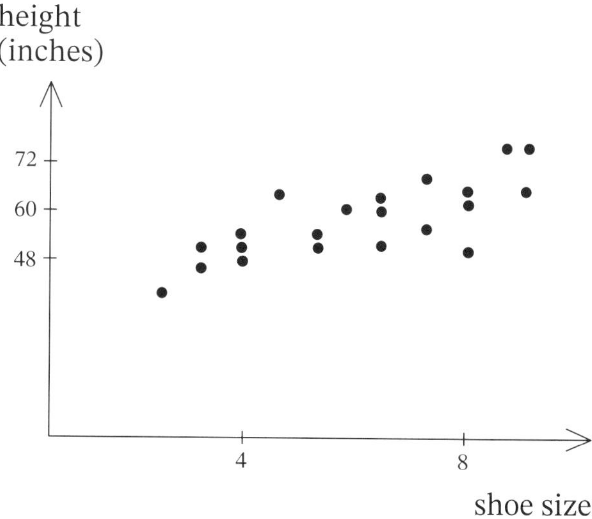

FIGURE 8

Scatter plots are intuitive and natural to work with. It is helpful to realize that *every* graph one draws in school mathematics classes is just a scatter plot. To draw a graph, one simply asks: What is the natural numerical data to collect from this situation?

Data from Equations

Recall that an equation is a statement that two mathematical expressions are equal. As such, each equation is a statement about numbers. For example, the equation $4 + 8 = 13 - 1$ is a true number sentence and the equation $9 + 1 = 20$ is a false one. Equations that involve variables are neither true nor false in and of themselves: their truth value as number sentences usually depends on the values selected for the variables. For example, the statement $b^2 = a^3 + 1$ is neither true nor false.

Comment. A *mathematical identity* is an equation that is true no matter what values are selected for the variables that appear in the equation. For example, the equation $n^2 - 1 = (n - 1)(n + 1)$ is a mathematical identity. (Warning: This definition is naive as there might be implied restrictions on the possible values one may select for the variable(s) that appear. For example, is $\frac{x^2}{x} = x$ a mathematical identity? The expression on the left is defined for all values of x except 0, but the expression on the right is valid for all values of x. Some textbook authors might not deem $x^2/x = x$ in and of itself as an identity because of this mismatch of allowable x-values. Other textbook authors do say it is one, arguing that it is implied that one's thinking should be restricted to those x-values that are allowable for both expressions.)

Given an equation, what is the natural data to collect from it? For example, what data could we collect from $b^2 = a^3 + 1$?

It seems natural to collect the set of all values a and b that make the equation a meaningful and true statement about numbers.

For example, $a = 2$ and $b = 3$ gives a true number statement: $(3)^2 = (2)^3 + 1$. We have one data value to put on our scatter plot. See Figure 9. We also have $a = 2$, $b = -3$; $a = 0$, $b = 1$; $a = 0$, $b = -1$; and $a = -1$, $b = 0$ as data points. Are there any more?

Curiously, these are the only integer values for both a and b that make the equation true. But there are certainly noninteger data points

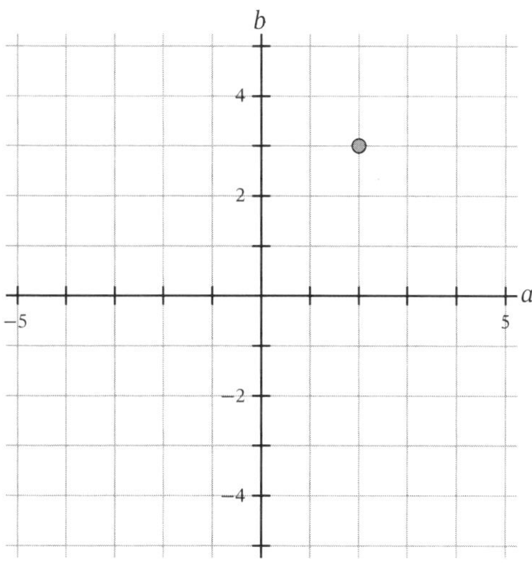

FIGURE 9

to plot as well. For instance, $a = 4$, $b = \sqrt{65} \approx 8.1$ and $a = \sqrt[3]{24} \approx 2.9$, $b = -5$ both make the equation true.

As one determines more and more data points that make the equation true, the scatter plot starts to outline the shape of a curve. Actually, it seems we have a whole continuum of data values to plot, and so we might well argue that the full scatter plot really is a curve rather than a collection of disconnected points. See Figure 10.

In summary:

> The *graph of an equation* is a scatter plot of all the data values that make the equation a meaningful and true statement about numbers.

One can have some quirky fun with this idea.

Exercise. Consider the equation $\dfrac{xy}{xy} = 1$. Describe its associated graph.

Solution. The equation $\dfrac{xy}{xy} = 1$ is meaningful and true whenever x and y each have a nonzero value. The equation is not well-defined if one or

both of these values is 0. Thus the graph of this equation is the set of all
points in the coordinate plane away from the axes.

□

FIGURE 10

Exercise. (a) What is the graph of the equation $\dfrac{\sqrt{x}\cdot\sqrt{y}}{\sqrt{x}\cdot\sqrt{y}} = 1$?

(b) What is the graph of $\left(\dfrac{2x}{|x|-x}+1\right)\left(\dfrac{2y}{|y|-y}-1\right) = 0$?

Graphs aren't always two-dimensional.

Example. The graph of the equation $x^2 + y^2 + z^2 = 1$ is a sphere in three-dimensional space.

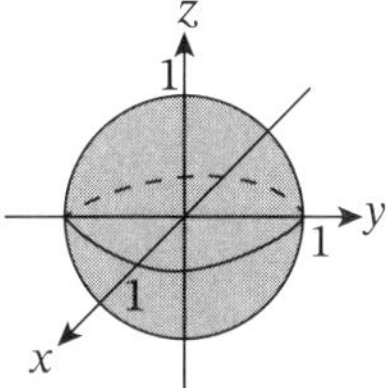

The graph of the equation $x^2 + y^2 = 1$ is a circle in two-dimensional space.

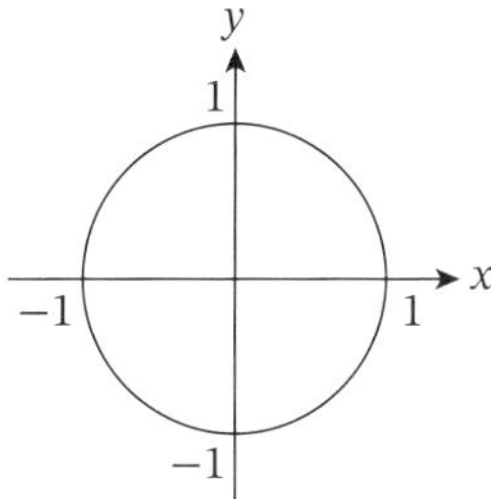

The graph of the equation $x^2 = 1$ is a circle in one-dimensional space.

Comment. Context, as always, is important! If $x^2 + y^2 = 1$ is actually an equation in three variables x, y, and z, but the variable z just doesn't happen to be mentioned (maybe the equation is $x^2 + y^2 + 0 \cdot z = 1$), then the graph of $x^2 + y^2 = 1$ is a cylinder sitting in three- dimensional space. (What is the graph of $x^2 = 1$ as an equation in two variables x and y? In three variables x, y, and z?)

Data from Functions

The natural data to collect from a numerical function is the set of all input/output pairs (p, q) with p an allowed input for the function and q its matching output.

For example, the function that assigns to each counting number the sum of its digits gives a scatter plot that looks like Figure 11. (We see the pairs $(7, 7)$ and $(35, 8)$, for instance, plotted.)

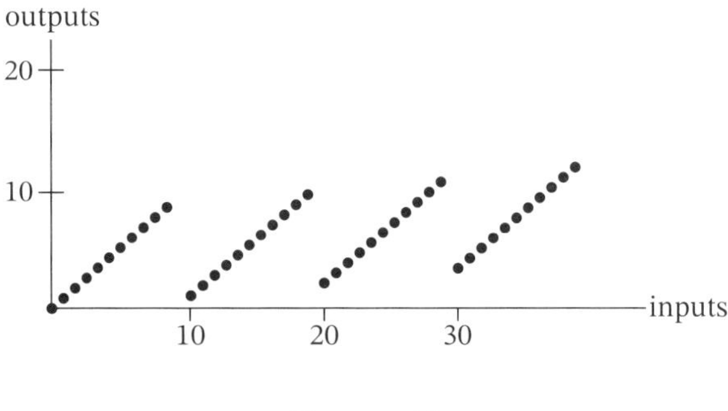

FIGURE 11

The squaring function, $s : \mathbb{R} \to \mathbb{R}$, which matches each real number with its square, gives a scatter plot that is a symmetrical U-shaped curve (Figure 12). Here we have the data points $(5, 25)$, $(0.1, 0.01)$, and $(-5, 25)$, for instance. Each data point has the form (x, x^2) and we can argue that the scatter plot is actually a whole continuum of data points forming a curve.

Comment. It is customary, when first plotting graphs, to plot just a few well-chosen data points to get a sense of the shape and behavior of the graph. (This is done in advanced mathematics too.) Many school curricula have students first collate preliminary data in a table.

In summary:

> The *graph of a numerical function* $f : \mathbb{R} \to \mathbb{R}$ is the scatter plot of all the data values (p, q) with p a valid input of the function and q its matching output.

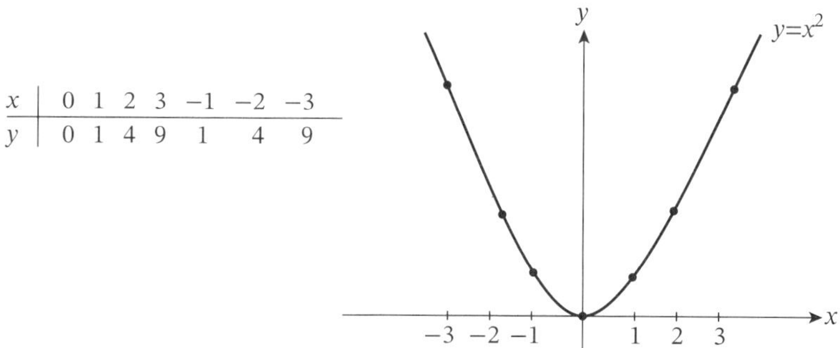

$$\begin{array}{c|ccccccc} x & 0 & 1 & 2 & 3 & -1 & -2 & -3 \\ \hline y & 0 & 1 & 4 & 9 & 1 & 4 & 9 \end{array}$$

FIGURE 12

Example. The *identity function*

$$f : \mathbb{R} \to \mathbb{R},$$
$$x \mapsto x$$

has graph

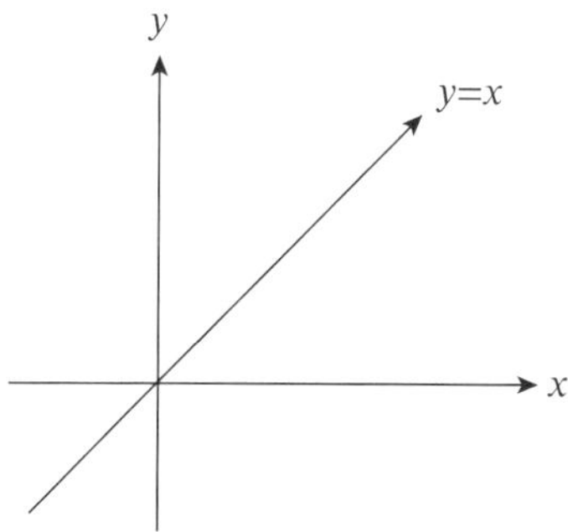

Example. Consider the constant function

$$f : \mathbb{R} \to \mathbb{R},$$
$$x \mapsto 23.$$

This function is given by the formula $f(x) = 23$, for all x. (So $f(1) = 23$, $f(-5) = 23$, and $f\left(256096\frac{1}{3}\right) = 23$, for example.) The graph of this function is

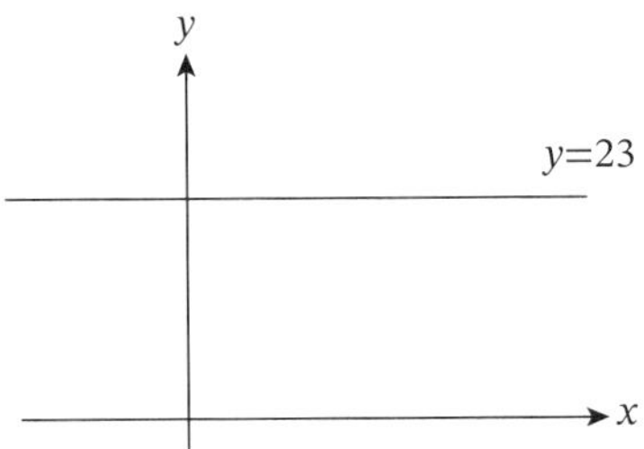

Notice here that the domain of this function is $\mathbb{R}$ and the range is $\{23\}$.

Example. Consider the following scatter plot. Is it the graph of a function?

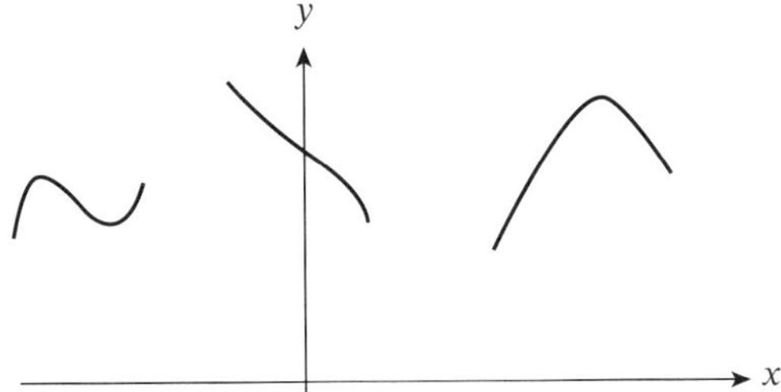

One can argue that it is the graph of a function, but one whose domain is just a subset of $\mathbb{R}$ (and this subset is composed of three intervals).

Exercise. If (p, q) is a data point for the function f, then q is the output matching the input p. We can also say that p is the input from which q came, and so (q, p) is a data point for the inverse function f^{-1}, if the inverse is meaningful.

How then does the graph of f^{-1} compare to the graph of f? (Must f^{-1} exist to play with this idea?)

Solution. The point (q, p) is the reflection of (p, q) across the northeast diagonal line through the origin of the Cartesian plane. Thus the graph of f^{-1} is the reflection of the graph of f across this diagonal line. See Figure 13. $\qquad\square$

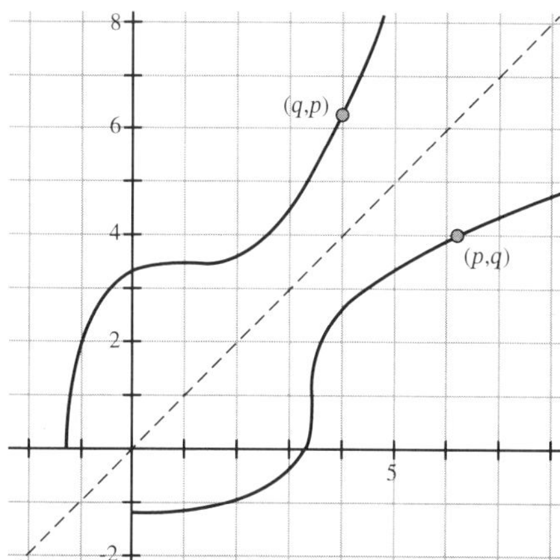

FIGURE 13

A Subtle Interplay

Recall that people often interpret an equation as a definition of a function. For example, the equation $y = x^2 + 1$ is sometimes interpreted as defining a function, let's name it g, by declaring $g(x) = x^2 + 1$. This means we potentially have two different graphs: the graph of the equation (the scatter plot of all the pairs of values that make the equation true) and the graph of the function (the scatter plot of all the input/output pairs).

Fortunately, if an equation is being interpreted as the definition of a function, then the two graphs are identical. In our example, for instance, if (x, y) is a data point on the scatter plot of the equation $y = x^2 + 1$, then x and y are values that make the equation true and so y equals $x^2 + 1$. Thus, thinking of the function g, y is the matching output for input x and so (x, y) is indeed a data point on the scatter plot for g. Conversely, if (p, q) is a data point on the scatter plot for g, then q is the matching output for input p and so $q = g(p) = p^2 + 1$. Thus (p, q) is a pair that makes the equation $y = x^2 + 1$ true and so (p, q) is on the scatter plot of the equation $y = x^2 + 1$.

In summary:

> Given a numerical function f, the graph of the equation $y = f(x)$ (the set of all pairs (x, y) that make this equation true) matches the graph of the function f (the set of all input/output pairs (x, y) for f).

Most people don't notice the subtleties at play here and will "just graph $y = f(x)$", not thinking of it one way or the other as the graph of an equation or as the graph of a function.

We saw earlier that if $f : \mathbb{R} \to \mathbb{R}$ has an inverse function, then the graph of the function f^{-1} is the reflection of the graph of the function f across a diagonal line.

> If (p, q) is on the scatter plot of f, then q is the output of f when applied to p. That is, p is the input when q came, and so p is the output of f^{-1} when applied to q. Thus (q, p) is a point on the scatter plot of f^{-1}

If one is thinking solely in terms of equations, then one can arrive at the same conclusion, but the reasoning is more subtle.

The equations $y = f(x)$ and $x = f^{-1}(y)$ are identical equations and so have the same scatter plot graphs. (For instance, $y = x + 2$ and $x = y - 2$ are equivalent equations.)

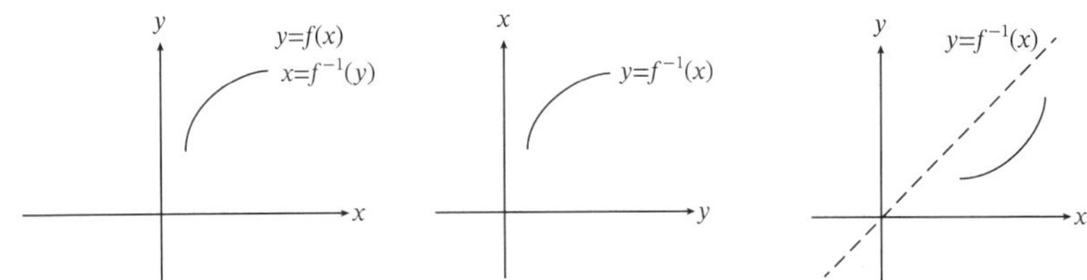

FIGURE 14

But in writing $x = f^{-1}(y)$ people usually prefer to use the symbol x for an input and y for an output and so will draw the graph of the equation with the x's and y's interchanged.

But this yields a diagram with x as a vertical axis and y a horizontal one. Society prefers the x-axis as horizontal and the y-axis as vertical, and one way to induce this change is to reflect the graph, axes and all, about a diagonal line.

Figure 14 shows three identical graphs of the inverse function, but only the third one meets the two societal conventions about the use of symbols.

Data from Sequences

What seems to be the natural data to collect from a sequence? For example, consider the sequence of Fibonacci numbers 1, 1, 2, 3, 5, 8, 13, ... What seems to be the natural data to collect and plot here?

Well, the first Fibonacci number is 1, the second is 1, the third is 2, and so on. This suggests we plot the data points $(1, 1)$, $(2, 1)$, $(3, 2)$, $(4, 3)$, $(5, 5)$, $(6, 8)$, and so on. This gives a visual presentation of the sequence. See Figure 15.

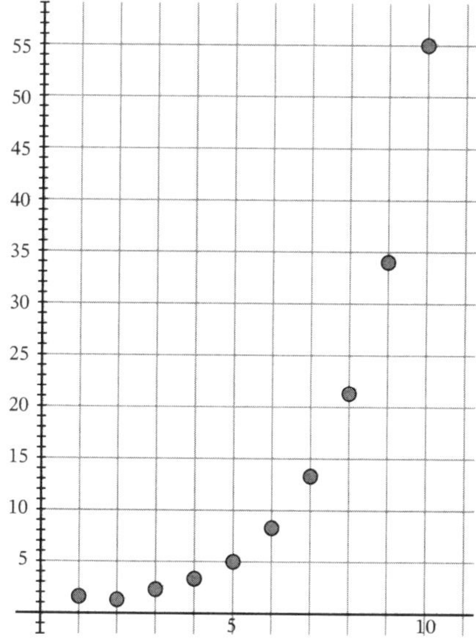

FIGURE 15

In summary:

> The *graph of a sequence* $\{a_n\}$ is the scatter plot of the data points (n, a_n).

If we think of a sequence as a function from $\mathbb{N}$ to $\mathbb{R}$, then the plot of the sequence matches the plot of the function.

Graphs Define Functions

Given a scatter plot, we can always construct a function from it. The domain of the function is the set of all values x on the horizontal axis that have at least one point vertically above it, and the output associated with x is the set of all second coordinates of points on the plot with first coordinate x. This gives a multi-valued function from $\mathbb{R}$ to subsets of $\mathbb{R}$.

For example, from Figure 16 we construct the function with domain $\{-3, -1, 1, 2, 3, 5\}$ given by

$$
\begin{aligned}
-3 &\mapsto \{6\}, \\
-1 &\mapsto \{-1, 4\}, \\
1 &\mapsto \{1, 4\}, \\
2 &\mapsto \{2, 5\}, \\
3 &\mapsto \{0\}, \\
5 &\mapsto \{1\}.
\end{aligned}
$$

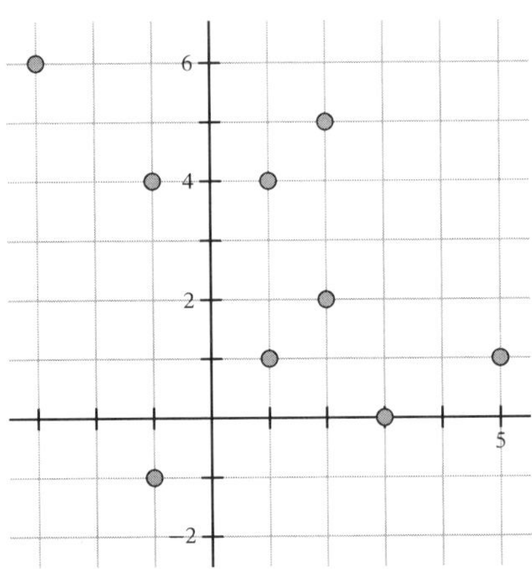

FIGURE 16

Most school text authors do not consider functions whose outputs are sets of numbers and will insist that a function have just one numerical output value associated with each input. Thus for a scatter plot to define a function in this sense it must satisfy a *vertical line test*: no two points in the scatter plot can have the same first coordinate (that is, no two points can sit in the same vertical line). For example, the scatter plot of Figure 17 defines a function f with one numerical value associated for each input. The domain of the function is, at a minimum, the interval of values $[-11, 11]$ (who knows what the diagram does beyond this interval?) and we see, for example, that $f(-10) = 2$ and $f(2) = 3.6$.

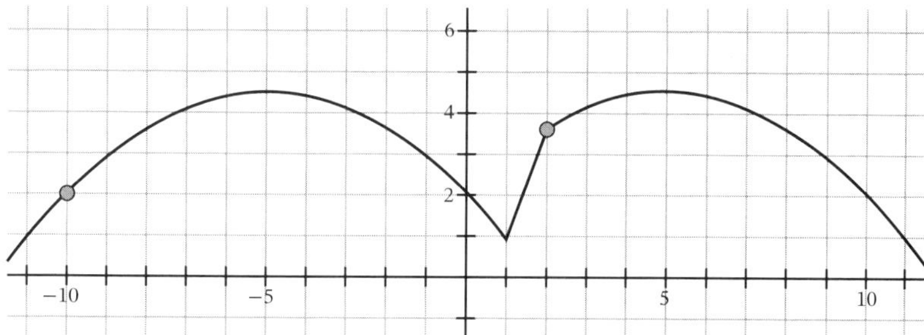

$$\textsc{Figure 17}$$

Exercise. The graph of a function passes the "horizontal line test". (No two points in the plot lie on the same horizontal line.) What does this indicate about the function?

Solution. For any point in the range of the function, there is just one input from which it came. Thus the function is one-to-one and it is possible to define an inverse function.   □

Challenge. Figure 18 shows, on the same set of axes, the graphs of two functions f and g, along with the sketch of a line of slope 1 though the origin. Nela is charged with sketching a graph of $y = g(f(x))$. For a selection of input values along the x-axis, she shall draw a series of vertical and horizontal line segments as shown to start identifying points on the graph of the composite function. Can you explain why her curious approach is valid?

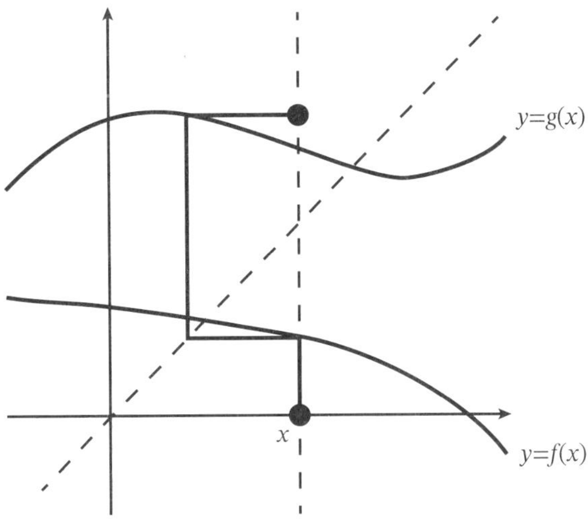

FIGURE 18

Jargon

There is standard jargon used to describe features of the graph of a function.

The *y-intercept* of a function is the location at which it crosses the (vertical) y-axis. For example, the y-intercept of the function shown in Figure 17 appears to be 2. (Some people insist the y-intercept be written as a point, $(0, 2)$, or as an equation, $y = 2$.)

The y-intercept of a function f is found by computing $f(0)$.

Comment. If the outputs of a function are not labeled "y", then the name *y-intercept* should be modified.

The *x-intercepts* of a function are the locations at which the function crosses the (horizontal) x-axis. For the function shown in Figure 17, there appear to be no x-intercepts in the domain of input values presented (though might $x = 12$ and $x = -12$ turn out to be x-intercepts?).

The x-intercepts of a function are the locations at which the function has output value 0. They can be found by examining the equation $f(x) = 0$. (And any solution to $f(x) = 0$ is called a *zero* or a *root* of f.)

Comment. Given an equation in two variables x and y, people also refer to the x- and y-intercepts of the equation. These are the locations at which the graph of the equation crosses the axes. One can (attempt to) find the location of the x-intercepts of the equation by setting $y = 0$ in the equation, and the y-intercepts by setting $x = 0$. (For example, the equation $(x - 4)^2 + (y - 10)^2 = 25$ has y-intercepts $y = 7$ and $y = 13$ found by solving $(-4)^2 + (y - 10)^2 = 25$. Attempting to solve $(x - 4)^2 + (-10)^2 = 25$ shows that it has no x-intercepts.)

A function f is said to be (strictly) *increasing* on an interval (a, b) if the outputs of the function increase in value as we examine larger and larger input values within the interval. (That is, if $a < x_1 < x_2 < b$, then we are sure to have $f(x_1) < f(x_2)$.) It looks like the function of Figure 17 is increasing on the intervals $(-11, -5)$ and $(1, 5)$.

We can also describe a function as (strictly) *decreasing* on an interval.

Loosely speaking, a function is said to have a *relative maximum* at a position $x = a$ if the value of $f(a)$ is larger than the output values $f(x)$ for all inputs x just to the left and right of a. (More precisely, $x = a$ provides a relative maximum for f if there is a number ϵ so that $f(a) > f(x)$ for all x in $(a - \epsilon, a + \epsilon)$.) Here we are assuming that the function is defined on a continuum of input values x.

It appears that the function f of Figure 17 has relative maxima at $x = -5$ and at $x = 5$. It also appears to have a *relative minimum* at $x = 1$.

One might also wish to describe over which range(s) of input values the function is positive or negative. There is no special language for these intervals.

Simultaneous Graphs

The graph of an equation involving two variables x and y is the scatter plot of all data points (x, y) of values that make the equation a true statement about numbers. Consequently, if we plot the graph of two equations on the same set of axes, then any point common to both plots (a *point of intersection*) is a pair of values that yields true number statements for each equation simultaneously.

Example. The equations $2x + y = 8$ and $3x - y = 7$ have simultaneous solution $x = 3, y = 2$. (Add the two equations together to see this.) Thus the graphs of the two equations must intersect at the point $(3, 2)$.

Sketching the graphs of these equations, it seems that this is the only point of intersection. In this case, $x = 3, y = 2$ must be the only simultaneous solution to the two equations.

Example. Graphing shows that the pair of equations $x^2 + y^2 = 1, x + 2y = 3$ has no simultaneous solution. See Figure 19.

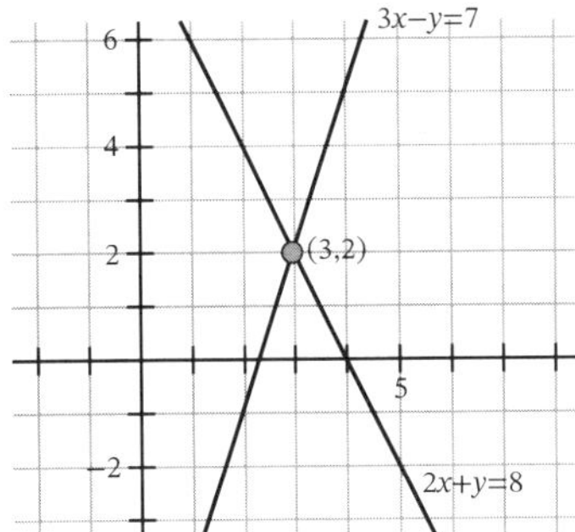

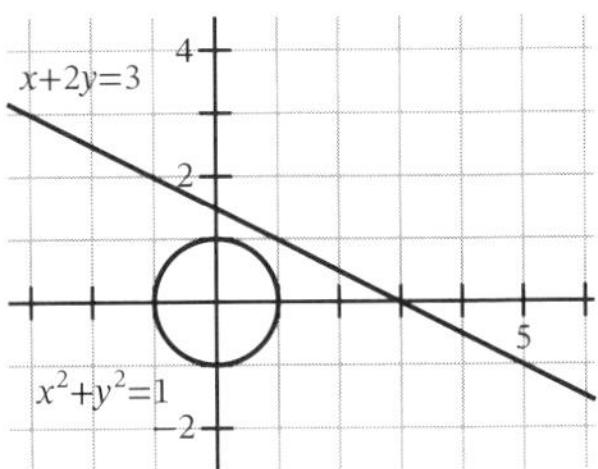

FIGURE 19

MAA Problems

26. (#8, AMC 10A, 2006) A parabola with equation $y = x^2 + bx + c$ passes through the points $(2, 3)$ and $(4, 3)$. What is c?

(A) 2

(B) 5

(C) 7

(D) 10

(E) 11

27. (#11, AMC 10A, 2006) Which of the following describes the graph of the equation $(x + y)^2 = x^2 + y^2$?

(A) The empty set

(B) One point

(C) Two lines

(D) A circle

(E) The entire plane

28. (#12, AMC 10B, 2006) The lines $x = \frac{1}{4}y + a$ and $y = \frac{1}{4}x + b$ intersect at the point $(1, 2)$. What is $a + b$?

(A) 0

(B) $\frac{3}{4}$

(C) 1

(D) 2

(E) $\frac{9}{4}$

29. (#20, AMC 12B, 2003) Part of the graph of $f(x) = ax^3 + bx^2 + cx + d$ is shown. What is b?

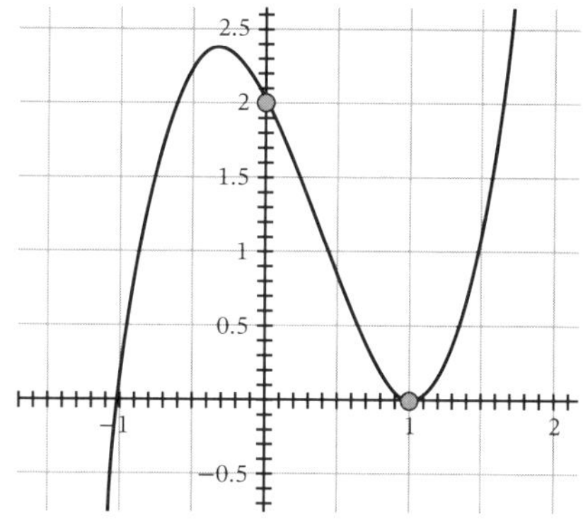

(This diagram is a redrawing of the original figure.)

(A) -4

(B) -2

(C) 0

(D) 2

(E) 4

6

Transformations of Graphs

Figure 20 shows the graph of a function $f : \mathbb{R} \to \mathbb{R}$. (Let's presume the graph continues to grow linearly to the left and to the right of the domain shown.)

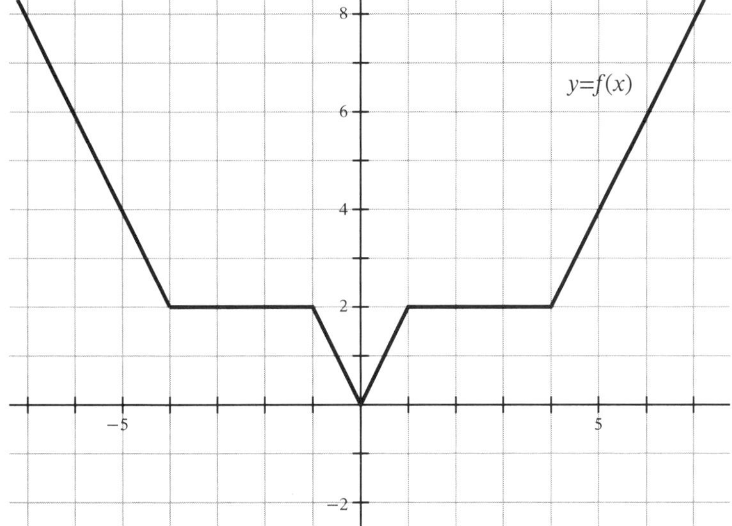

FIGURE 20

We see here that $f(4) = 2$, $f(-5) = 4$, and $f(0) = 0$, for instance. For each real number input x there is an output value $f(x)$, which can be read from the graph even though we don't know a specific formula for that output. We can equivalently regard Figure 20 as the graph of the

71

equation $y = f(x)$. (Recall that the graph of a function f matches the graph of the equation $y = f(x)$.)

Now let's consider ten functions created by modifying the outputs of f. In particular, let's consider the functions that associate with each real number x the output

$$f(x - 2), \ f(x + 2), \ f(x) - 2, \ f(x) + 2,$$

$$f(2x), \ f\left(\frac{1}{2}x\right),$$

$$2f(x), \ \frac{1}{2}f(x),$$

$$f(-x),$$

and

$$-f(x),$$

in turn. (This is a standard thing to study in a school curriculum.) We'll see how to deduce what their matching graphs must be. We'll also offer some brief explanation as to why one might want to modify graphs of functions these ways.

The Graph of $y = f(x - 2)$

The value 2 is prominent in this equation. Notice that putting $x = 2$ into $f(x - 2)$ gives $f(0)$. So, in some sense, the value 2 is now "behaving like 0" for the function f. Consequently, whatever the graph of $y = f(x)$ does at $x = 0$, we deduce that the graph of $y = f(x-2)$ is doing the same thing, but now at $x = 2$. This suggests that the graph of $y = f(x - 2)$ appears as in Figure 21.

We can check that this is indeed correct by examining other values: for $x = 6$ the value of $f(x-2)$ is $f(4) = 2$ and this graph does indeed have value 2 at $x = 6$; at $x = -3$ this graph matches the value of $f(-3 - 2) = f(-5) = 4$.

The graph of $y = f(x + 2)$

Here $x = -2$ is behaving like 0. This suggests $f = f(x+2)$ has the graph in Figure 22.

At $x = 2$, $f(x + 2)$ has value $f(4) = 2$, and one can check that this graph is correct for other input/output pairs.

In summary:

The graph of $y = f(x-k)$ is a horizontal translation of the graph of $y = f(x)$ with k now behaving like 0.

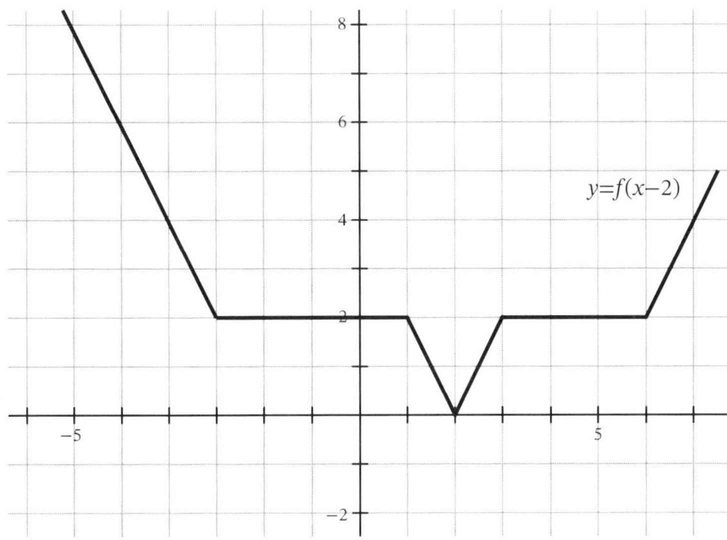

FIGURE 21

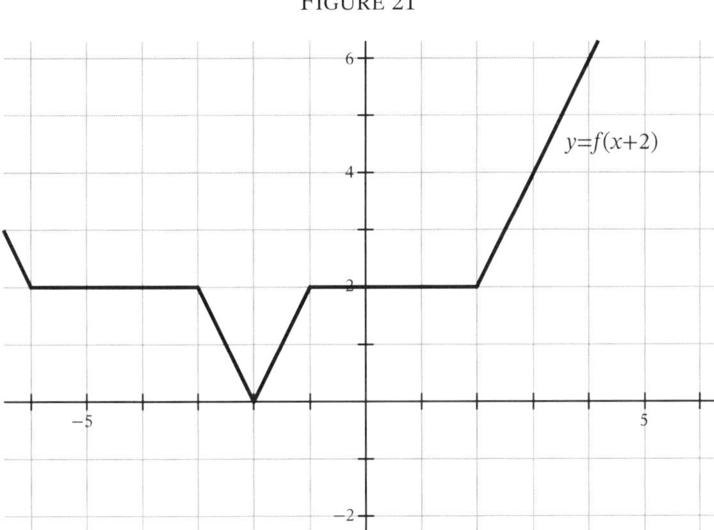

FIGURE 22

There is no need to memorize if the translation is a left or right horizontal shift. Just ask: "Which value of x is now behaving like 0?"

If one is collecting data for a scientific experiment and later finds that an instrument was not correctly calibrated, say a timer was off by two milliseconds, then one might want to adjust all the data values $f(t)$ to $f(t-2)$.

The graph of $y = f(x) - 2$

Here we see that each output value $f(x)$ is being decreased by 2. Thus the graph of $y = f(x) - 2$ is the same as the graph of $y = f(x)$ but with each second coordinate decreased by 2. (See Figure 23.) For example, at $x = -5$, the value of $f(x) - 2$ is $f(-5) - 2 = 4 - 2 = 2$.

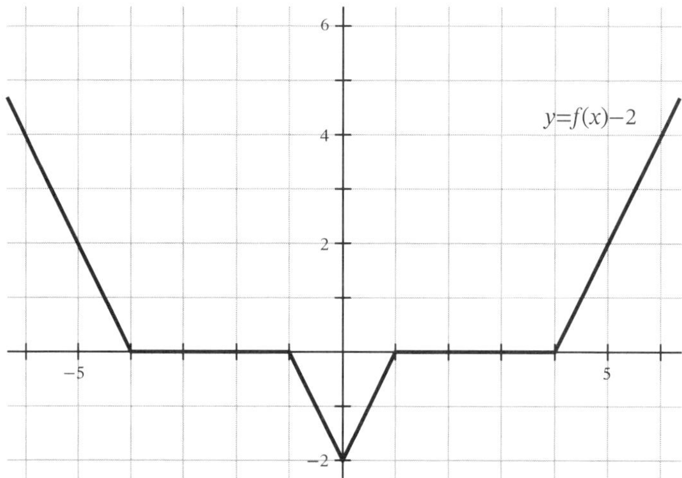

FIGURE 23

The graph of $y = f(x) + 2$

Here each output value is increased by 2. The graph of $y = f(x) + 2$ is thus an upward translation of the graph of $y = f(x)$ by two units.

In summary:

The graph of $y = f(x) + k$ is a vertical translation of the graph of $y = f(x)$ by k units.

(If you like, you can think of $y - k = f(x)$ as stating that k is the new zero of the y-values: whatever the original graph was doing at $y = 0$, the new graph does at $y = k$.)

Again, in a scientific experiment, if an instrument recording output values was not properly calibrated, then one might well want to adjust the output readings by a fixed value.

Example. The equation of a circle of radius r, center $(0,0)$, is $x^2 + y^2 = r^2$. Then $(x-2)^2 + (y+5)^2 = r^2$ is the same equation, but with 2 behaving like 0 for the x-values and -5 behaving like 0 for the y-values. Thus it is the equation of a circle of radius r, center $(2,-5)$.

Example. The absolute value function $d : \mathbb{R} \to \mathbb{R}$ assigns to each real number its distance from the origin on the number line. For example, since the number 2 on the number line is two units from the origin, we have $d(2) = 2$, and with -7 seven units from the origin, $d(-7) = 7$. Also, $d(0) = 0$. In general,

$$d(x) = \begin{cases} x & \text{if } x \text{ is nonnegative,} \\ -x & \text{if } x \text{ is negative.} \end{cases}$$

Figure 24 shows the graph of this function. It is a V-shaped graph with vertex at the origin.

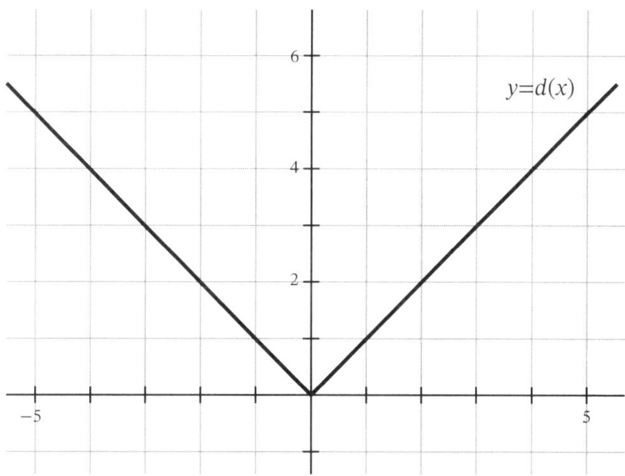

FIGURE 24

Exercise. I am six units tall and am standing at the position $x = 4$ on the horizontal axis. Write a defining formula for a function whose graph is the same V-shape as the absolute value graph, but positioned so as to be balancing on my head. (See Figure 25.)

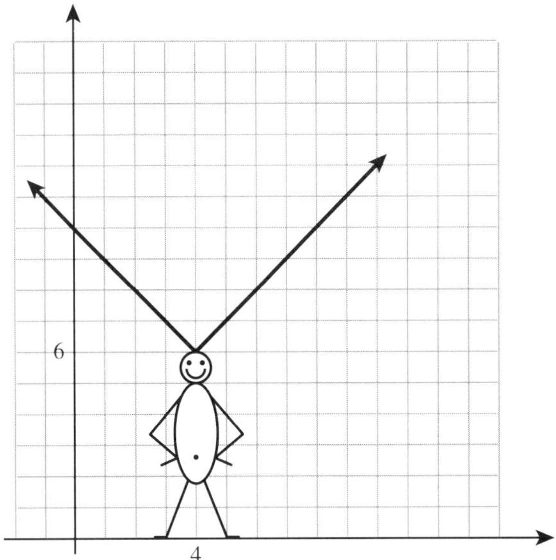

FIGURE 25

Solution. We want $x = 4$ to be the new zero of the x-values and all the output values to be increased 6 units. Thus use $y = d(x - 4) + 6$.   □

Comment. The absolute value of a number x is usually denoted $|x|$ or $\text{abs}(x)$.

Exercise. Describe the graph of $y = |x + 5| - 5$.

Solution. This is a V-shaped graph with vertex $(-5, -5)$. (We see that $x = -5$ is behaving like 0 for the x-values and that the outputs are all decreased by five units.)   □

The graphs of $y = f(2x)$ and $y = f\left(\frac{1}{2}x\right)$

Returning to the graph of $y = f(x)$ of Figure 20, let's consider the graph of $y = f(2x)$.

Here we run through the x-values at twice the speed. When $x = 1$, we have $y = f(2 \times 1) = f(2)$; when $x = 3$, we have $y = f(6)$; and when $x = -2\frac{1}{2}$, we have $y = f(-5)$. But we still have $y = f(0)$ for $x = 0$. The graph of $y = f(x)$ is essentially the same as the graph of $y = f(x)$ except that it will whip through the output values twice as quickly to the left and to the right of $x = 0$. See Figure 26.

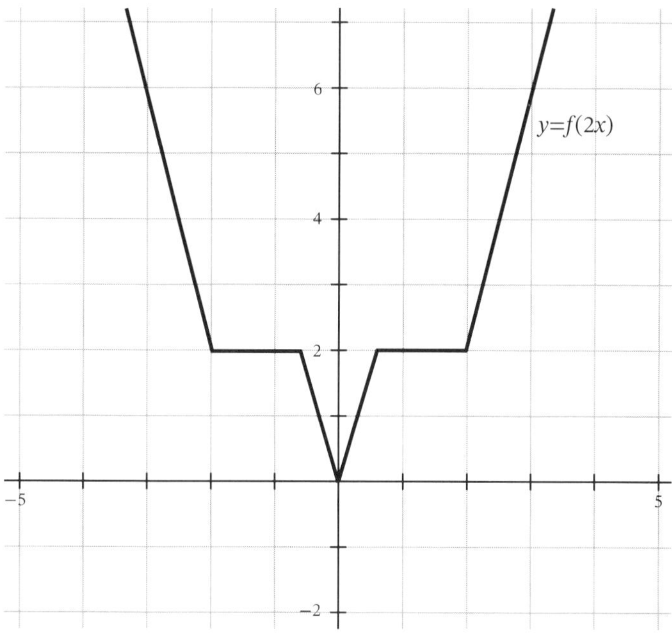

FIGURE 26

In the same way, the graph of $y = f(10x)$ will run through outputs ten times as quickly (when $x = 5$, for example, we are already working with the output value $f(50)$) and the graph of $y = f\left(\frac{1}{2}x\right)$ will do so at half the speed. (When $x = 4$, we are still only working with the output value $f(2)$.) See Figure 27.

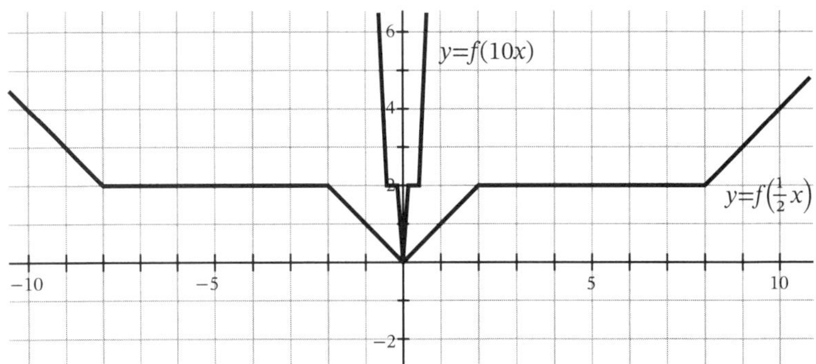

FIGURE 27

In summary:

The graph of $y = f(kx)$ for a positive value k changes the input values so that the graph runs through the output values k times as fast. It has the visual effect of horizontally compressing (if $k > 1$) or stretching (if $0 < k < 1$) the graph about $x = 0$.

If one is conducting an experiment and the rate of a running instrument is not calibrated to the correct speed, say, there is a rate of error of 1%, then one might well wish to adjust the recorded data values by a factor of 1.01.

If $y = f(x)$ records outputs for inputs x measured in centimeters, then $y = f(2.54x)$ records the data with inputs measured in inches.

A sine curve $y = \sin(x)$ is often used to model periodic phenomena. The equation $y = \sin\left(\frac{2\pi}{T}x\right)$ has $x = T$ behaving like $x = 2\pi$ and so gives a graph of a sine curve of a different period. (See *Trigonometry: A Clever Study Guide* for more on this.[1]

Example. Consider the equation $y = f(2x - 6)$. We can see two possible effects at play here: some new value of x is behaving like 0, and we're running though x-values at twice the speed. How do we think through the combination of these two effects?

The answer is to imagine building this equation in stages and looking at the graphs of the equations we build in turn.

---

[1] James Tanton, *Trigonometry: A Clever Study Guide*, Volume 25, AMS/MAA Press, 2015, https://bookstore.ams.org/prb-25/.

We have the graph of $y = f(x)$ from Figure 20. The graph of $y = f(2x)$ is that of Figure 26. The graph of $y = f(2x - 6)$ does have a new x-value behaving as 0. We see that $x = 3$ makes $2x - 6$ equal to 0. Thus the graph of $y = f(2x - 6)$ must appear as in Figure 28.

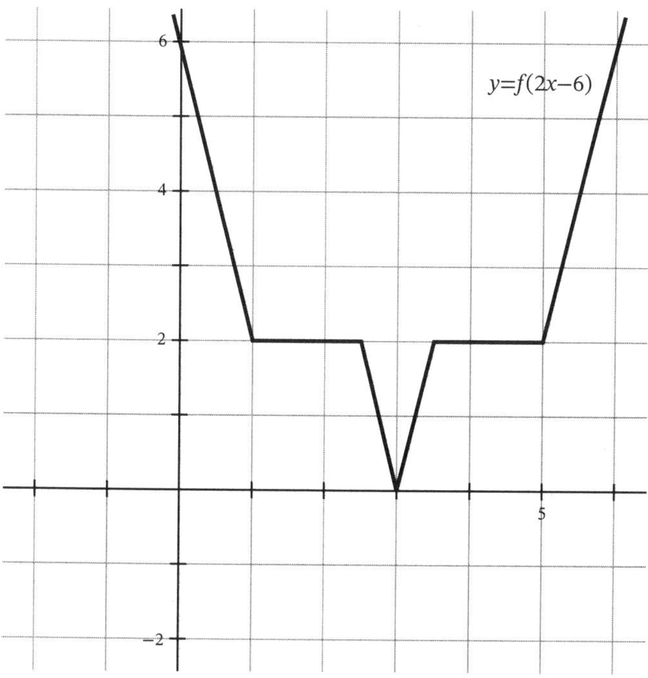

FIGURE 28

Alternatively, we can work with $y = f(x - 6)$ whose graph is a horizontal translation with $x = 6$ behaving as 0. Then $y = f(2x - 6)$ is the same graph compressed by a factor of 2, and so now $x = 3$ is behaving like 0.

Question. Can we instead work with $y = f(2(x - 3))$? It is clear here that $x = 3$ is now behaving as 0, which is good to see. But how do we interpret the effect of multiplying the term $x - 3$ by 2? Can we think through what it means to "run through the $x - 3$ values twice as fast"?

In general, the graph of $y = f(ax + b)$ with a a positive number matches the graph of $y = f(x)$, except that it runs through the x-values

a times as fast and has $x = -\dfrac{b}{a}$ behaving like 0. The graph of $y = f(x)$ is thus compressed or stretched on either side of 0 and then shifted horizontally $-\dfrac{b}{a}$ units to create the graph of $y = f(ax + b)$.

Exercise. Describe the graph of $y = f(4(x - 3) + 10)$.

Solution. It is hard to unravel three layers of thinking here. But we can see that we will have a speed effect induced by a term $4x$ and that there is a new value of x behaving like 0: from $4(x - 3) + 10 = 0$ we get $x = \dfrac{1}{2}$. Thus the graph in question is the graph of $y = f(x)$ compressed left and right by a factor of 4 and then shifted half a unit to the right (so that $x = \dfrac{1}{2}$ is behaving like 0). $\qquad\square$

Exercise. Describe the graph of $y = f\left(\dfrac{x+3}{2}\right)$.

Solution. This is the graph of $y = f(x)$ stretched horizontally according to a factor of $\dfrac{1}{2}$ and then shifted three units so that $x = -3$ behaves like 0. $\qquad\square$

Example. Let's sketch the graph of $y = \left|\dfrac{2x-10}{3}\right| - 1$.

We know this graph matches the V-shaped graph of $y = |x|$ stretched and translated, so it too will be a V-shaped graph. We see that it is a graph running through the x-values at $\dfrac{2}{3}$ the speed, with $x = 5$ behaving like 0, and with all outputs decreased 1 unit. We can say, right away, that the vertex of the graph will be at $x = 5$. In fact, it has coordinates $(5, -1)$.

We need now the slopes of the two arms of the graph. These are $\dfrac{2}{3}$ and $-\dfrac{2}{3}$, but they are awkward numbers to work with. Perhaps the easiest thing to do is to determine the coordinates of one more point on the graph and use the symmetry of the V-shaped graph to guide us.

I see that $x = 2$ gives an integer output, $y = 1$.

We can now draw the graph of this equation with ease. See Figure 29.

Exercise. Describe the graph of $y = \left|\dfrac{7x-5}{18} + 2\right| + 9$.

Solution. This is a V-shaped graph. The new zero for the x-values is given by $\dfrac{7x-5}{18} + 2 = 0$, and so is $x = -\dfrac{31}{7}$. The vertex of the graph is thus

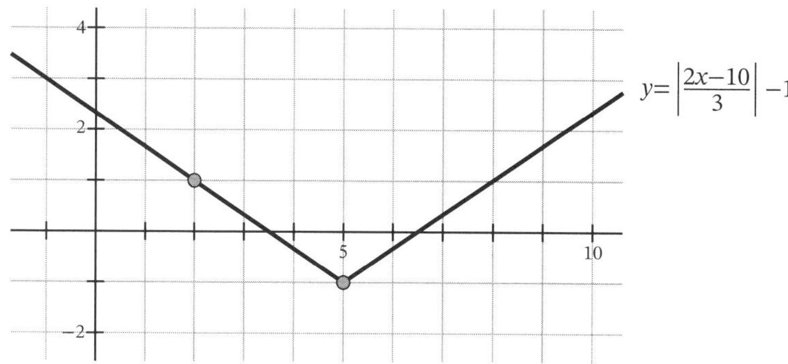

FIGURE 29

$\left(-\frac{31}{7},9\right)$. The graph also passes through the point $(11,15)$. This, with symmetry, provides enough information to sketch the graph. $\qquad\square$

Comment. The graph of $y = x^2$ is a symmetrical U-shaped graph with vertex at the origin. One can use similar reasoning to sketch a graph of $y = \left(\frac{7x-5}{18} + 2\right)^2 + 9$.

The graphs of $y = 2f(x)$ and $y = \frac{1}{2}f(x)$

The graph of $y = 2f(x)$ matches the graph of $y = f(x)$ except all the output values are doubled: an output value of 2 is now recorded as an output value of 4, an output value of 6 as a value of 12, and so on. (If there were negative outputs, they would be doubled too.) This has the effect of stretching the graph vertically either side of the horizontal axis. The graph of $y = \frac{1}{2}f(x)$ has all the outputs halved in value. See Figure 30.

In summary:

> *The graph of $y = kf(x)$ for k a positive number matches the graph of $y = f(x)$ but with all the output values adjusted by a factor of k. This has the visual effect of stretching, or compressing, the graph upwards above the horizontal axis and downwards below the horizontal axis.*

The curve $y = \sin(x)$ is often used to model periodic phenomena. The graph of $y = A\sin(x)$ oscillates between the values A and $-A$,

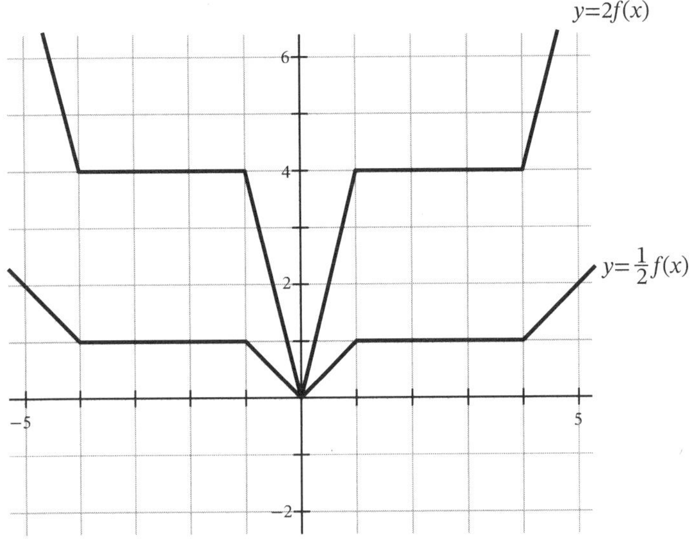

FIGURE 30

where A is the amplitude of the phenomenon being studied. (See *Trigonometry: A Clever Study Guide* for more on this. Refer to the footnote on page 78.)

Example. The graph of $y = 3f(2x)$ has a horizontal compression (we are running though the x-values twice as fast) and a vertical stretch (all outputs are tripled). See Figure 31.

Exercise. Describe the graph of $y = \frac{1}{2}f(2x - 5) + 3$.

Solution. There is a lot going on here.

One advantage of Euler's function notation is that we can follow the standard order of operations from algebra to work our way through matters: work with the innermost parentheses first, and work with multiplications and divisions before additions and subtractions.

Focus first on "$2x - 5$" and, within that, the "$2x$" term. Then move outwards from there. See Figure 32.

The graph of $y = f(2x)$ is the same as the graph of $y = f(x)$ but running through the x-values twice as fast.

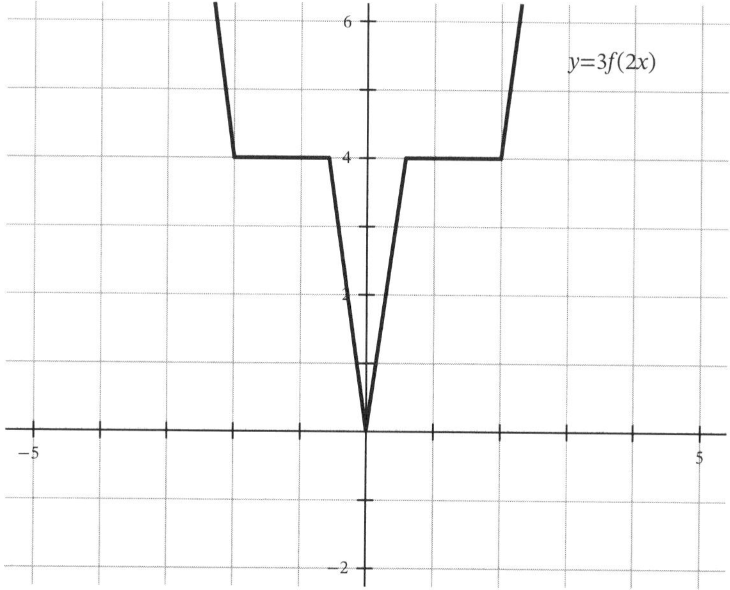

FIGURE 31

The graph of $y = f(2x - 5)$ is the same as the previous one but now with $x = 2\frac{1}{2}$ the new zero. We have a horizontal shift.

The graph of $y = \frac{1}{2}f(2x - 5)$ is the same as the previous one but with all outputs halved.

The graph of $y = \frac{1}{2}f(2x - 5) + 3$ is the same as the previous one but with all outputs increased by 3.

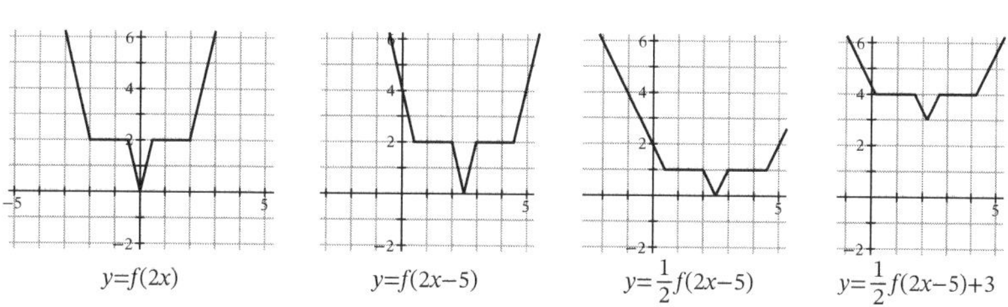

FIGURE 32

One can (and should!) check that this is correct by computing a few outputs. At $x = 1$ our final graph should indicate the output

$$\frac{1}{2}f(2 \cdot 1 - 5) + 3 = \frac{1}{2}f(-3) + 3 = \frac{1}{2} \cdot 2 + 3 = 1,$$

which it does, and at $x = 0$ the graph should indicate the output

$$\frac{1}{2}f(-5) + 3 = \frac{1}{2} \cdot 4 + 3 = 5,$$

also good! $\square$

The graphs of $y = f(-x)$ and $y = -f(x)$

In the equation $y = f(-x)$ the sign of each input is changed. Thus with the input $x = 3$, say, we associate the output of $x = -3$, and vice versa. This has, in effect, switched the roles of the positive and negative inputs on the horizontal axis of the graph and so produces a graph that is the mirror image of the original graph about the vertical axis.

Example. The graph of $y = f(x)$ in our example of Figure 20 is symmetrical about the vertical axis. Thus the graph of $y = f(-x)$ is identical to this graph.

Example. The graph of $y = \frac{1}{2}f(2(-x) - 5) + 3$ is the reflection of the graph of $y = \frac{1}{2}f(2x - 5) + 3$ about the vertical axis. (See Figure 33.)

Comment. To help parse this example, let's give $\frac{1}{2}f(2x - 5) + 3$ the name $g(x)$. Then Figure 32 shows the graph of $y = g(x)$ and Figure 33 shows the graph of $y = g(-x)$.

In the equation $y = -f(x)$ the sign of each output is changed: any positive output is made negative, and vice versa. This has the effect of reflecting the original graph across the horizontal axis.

Example. The graph of $y = -\left(\frac{1}{2}f(2(-x) - 5) + 3\right)$ is the reflection of the previous graph about the horizontal axis. (See Figure 34.)

Comment. If it helps, let $h(x)$ be $\frac{1}{2}f(2(-x) - 5) + 3$. Then Figure 33 is the graph of $y = h(x)$ and Figure 34 is the graph of $y = -h(x)$.

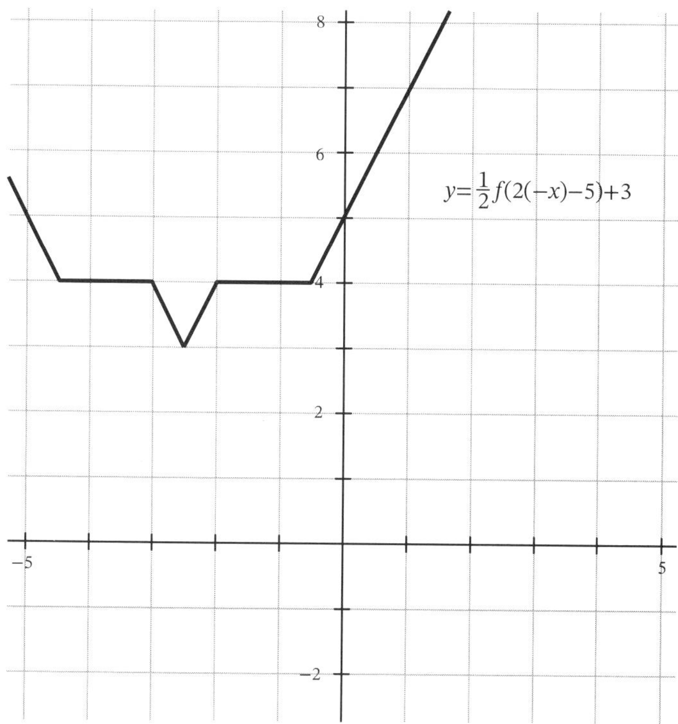

FIGURE 33

Here is an interesting sequence of ideas.

The graph of $y = x$ is a line of slope 1 through the origin.

The graph of $y = 3x$ is the same graph running through the x-values three times as fast. It is thus a line through the origin that rises three times as fast—it has slope 3.

The graph of $y = 3x + 7$ is the same as the previous graph except all outputs are increased by 7 units. Thus the graph is a line of slope 3 crossing the vertical axis at 7.

In general, our transformational thinking shows that the graph of $y = mx + b$ must be that of a straight line of slope m with vertical intercept b.

Exercise. Describe the graph of $y = 2 - |4x - 3|$.

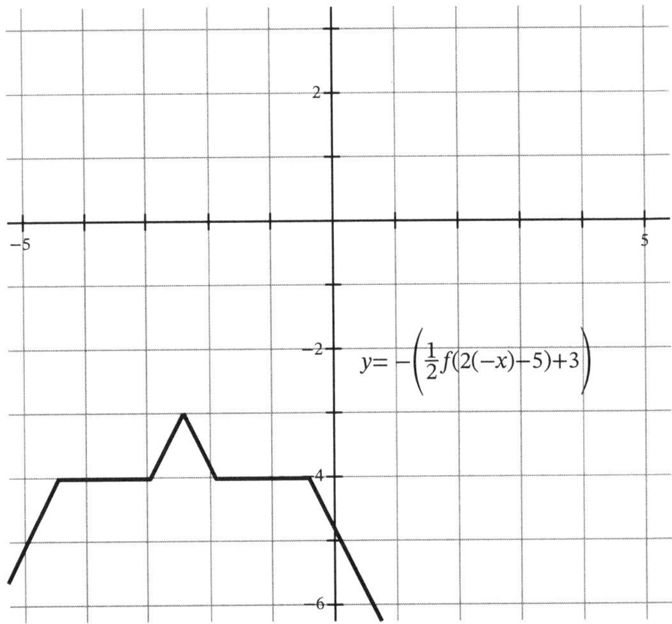

FIGURE 34

Solution. This is an upside-down V-shaped graph with vertex $\left(\frac{3}{4},2\right)$ and arms of slope 4 and -4. $\qquad\square$

Exercise. The function of Figure 20 has $f(x) = f(-x)$ for all real inputs x. Is there a function with $f(x) = -f(x)$ for all x? How about one with $f(x) = -f(-x)$ for all inputs x?

Solution. If $a = -a$, then a must be 0. Thus the constant function $f(x) = 0$ for all inputs x is the only function satisfying the first condition. There are many examples of functions satisfying the second condition with $f(x) = x^3 + 2x$ being one such example. $\qquad\square$

A function satisfying $f(-x) = f(x)$ for all inputs x has a graph symmetrical about the vertical axis and is called an *even* function. Symmetry plays an important role in many physical applications and it has become convenient to have some jargon describing these functions.

A function satisfying $f(-x) = -f(x)$ for all inputs (this is the second condition of the previous exercises) is called an *odd* function. The graph of $y = -f(-x)$ is obtained from the graph of $y = f(x)$ by performing a reflection about the horizontal axis followed by a reflection about the vertical axis. This is equivalent to preforming a 180° rotation about the origin. Thus the graph of an odd function has 180° rotational symmetry about the origin.

One can test whether a function is even or odd or neither by examining the symmetry of its graph or by algebraic manipulation: Is the value of $f(-x)$ sure to match the value of either $f(x)$ or $-f(x)$ for each and every possible value of x?

Example. The function g given by $g(x) = x^4 + x^2 + 1$ is even as $g(-x) = (-x)^4 + (-x)^2 + 1 = x^4 + x^2 + 1 = g(x)$ for all x. The function k given by $k(x) = x^3 + 2x$ is odd as one can see that $k(-x)$ equals $-(x^3 + 2x) = -k(x)$ for all x. The function q given by $q(x) = x^3 + x^2$ is neither even nor odd.

Exercise. For which integers n is the function f given by $f(x) = x^n$ even? For which n is it odd?

Solution. We have that f is even for all even values of n (including $n = 0$) and odd for all odd values of n. $\qquad\square$

Comment. It is not always easy to determine whether or not a function is even or odd or neither by algebraic methods. For example, the function given in Figure 20 is clearly even, yet the formula used to produce the graph is

$$f(x) = |x + 1 - |x - 1|| + |x - 4| + |x + 4| - |x + 1| + x - 7.$$

Are you able to show algebraically that $f(-x) = f(x)$ for all x?

Exercise. The graph of w given by $w(x) = |x - 3| - 3$ is a V-shaped graph with vertex at $(3, -3)$. Sketch the graphs of $y = w(|x|)$ and $y = w(-|x|)$.

Solution. The graph of the first equation is identical to the graph of $y = w(x)$ to the right of the vertical axis, and to the left it is a reflection of this right half. (The graph thus looks like a "W".) The graph of the second equation is identical to the graph of $y = w(x)$ to the left of the vertical axis, and to the right it is a reflection of this. (The graph, surprisingly, is thus identical to the graph of $y = |x|$.)

(As an additional challenge, describe the graph of $y = |w(x)|$.) $\quad\square$

Sometimes even and odd functions are seen as the building blocks of all other functions. Let f be a function and set $e(x) = \frac{f(x)+f(-x)}{2}$ and $o(x) = \frac{f(x)-f(-x)}{2}$. Then one can see that $e(-x) = e(x)$ and $o(-x) = -o(x)$ and so e and o are, respectively, even and odd functions. One can also see that $f(x) = e(x) + o(x)$. Thus every function is the sum of an even function and an odd function. (In fact, this decomposition is unique: f is the sum of no other even and odd function. Can you prove this is so?)

MAA Problems

30. (#4, AMC 10A, 2004) What is the value of x if $|x - 1| = |x - 2|$?

(A) $-\dfrac{1}{2}$

(B) $\dfrac{1}{2}$

(C) 1

(D) $\dfrac{3}{2}$

(E) 2

31. (#7, AMC 12B, 2005) What is the area enclosed by the graph of $|3x| + |4y| = 12$?

(A) 6

(B) 12

(C) 16

(D) 24

(E) 25

7

Average Rate of Change, Constant Rate of Change

We humans tend to be linear thinkers. On a road trip, if we traveled 360 miles in six hours, then we feel it is meaningful to say that we traveled an average of 60 miles per hour, even though we know full well we did not travel that speed consistently—we stopped for rests, we sped, we got caught up in traffic, and so on. The stock market is erratic, full of ups and downs, but it seems helpful to us to have a sense of the average rate of growth of the market from year to year.

We obtain average rates by looking at just two data points and ignoring all details of the other data points between them. For the road trip example, we observed that at start time, call it 0 hours, we had traveled 0 miles, and at end time, 6 hours, we had traveled 360 miles. That makes for an average speed of

$$\frac{360 - 0}{6 - 0} = 60 \text{ miles per hour.}$$

On January 14, 2000, the Dow Jones Index had value 11,722.98 and on January 15, 2016, its value was 16,466.30. That's an average growth rate of

$$\frac{16466.30 - 11722.98}{2016 - 2000} \approx 296.46 \text{ points per year}$$

for that time period.

Whether or not such figures are meaningful is to be debated. For example, the Dow Jones Index had increased by another 2100 points by the summer of 2016.

Consider a plot of quantity values over a period of time. Then, mathematically, the average rate of change of that quantity over the period between two time values is the slope of the line segment connecting the two data points for those time values. Again, this analysis ignores all other data values. See Figure 35.

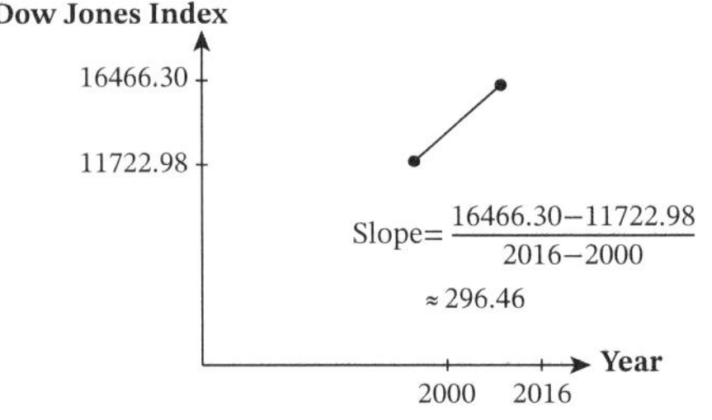

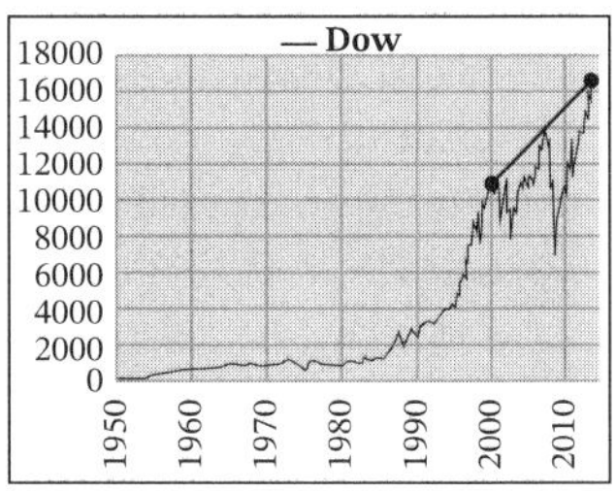

FIGURE 35

A graph is a plot of data points and so we can generalize this notion to arbitrary functions. If f is a function, then the *average rate of change* of f from input $x = a$ to input $x = b$ is the ratio of the change of function values compared to the change in input values: $\dfrac{f(b)-f(a)}{b-a}$. This matches the slope of the line segment connecting the two data points $(a, f(a))$ and $(b, f(b))$. See Figure 36.

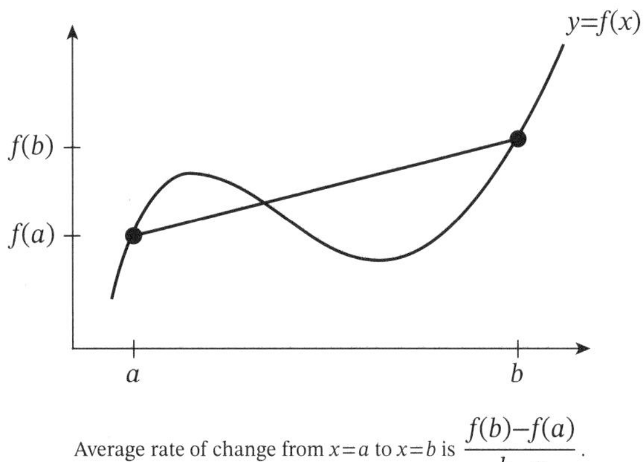

Average rate of change from $x=a$ to $x=b$ is $\dfrac{f(b)-f(a)}{b-a}$.

FIGURE 36

We usually associate the word "rate" with a sense of time—a change of value over a change of time—and so it can feel somewhat strange to use the word "rate" for nontime based examples. But such examples can naturally occur. If I always give 10% of my earned income to charity, then I am donating at a rate of ten cents per dollar. If my expected length of life increases by three weeks for every two pounds of weight I lose over 200 pounds, then I am gaining life expectancy at a rate of one-and-a-half weeks per pound until my weight is 200 pounds.

Exercise. Consider the squaring function s given by $s(x) = x^2$. Show that the average rate of change between consecutive whole square numbers grows without bound.

Solution. The average rate of change between the nth and $(n + 1)$th square numbers is

$$\frac{s(n + 1) - s(n)}{(n + 1) - n} = \frac{(n + 1)^2 - n^2}{1} = 2n + 1.$$

This value grows without bound as n gets larger and larger. $\qquad\square$

Exercise. A function F has the property that the average rate of change of the function always evaluates to $-\frac{4}{3}$ no matter over which two input values it is computed. What can you say about the function?

Solution. Consider one fixed input value a and a general input value x. We have that $\frac{F(x) - F(a)}{x - a} = -\frac{4}{3}$ for all values x. This tells us that $F(x) = -\frac{4}{3}(x - a) + F(a) = mx + b$ with $m = -\frac{4}{3}$ and b is the fixed value $F(a) - \frac{4}{3}a$. Thus F is a linear function. (And this makes intuitive sense too: if the slope of a line segment connecting any two points on the graph of F is constant, it seems that all the points of the graph should be collinear.) $\qquad\square$

Constant Rate of Change

Recall that in the standard school curriculum a function $f : \mathbb{R} \to \mathbb{R}$ is called a *linear function* if its graph is a straight line. But what makes a line straight?

One might loosely define a line plotted on a set of axes to be "straight" if the angle of elevation of the line, with respect to the horizontal, is the same at all points along the line. See Figure 37.

This means all right triangles drawn under pairs of points along the line are similar by the AA (angle angle) similarity principle. Consequently, all ratios of "rise over run" are the same for all such right triangles. The fixed ratio value rise over run is usually called the slope of the line.

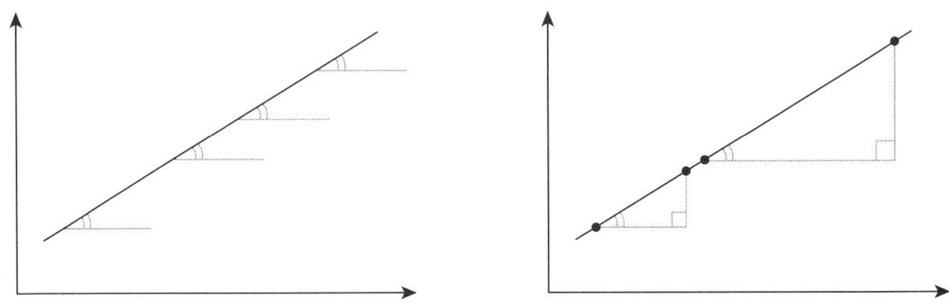

FIGURE 37

For our linear function f, the right triangle under the points $(a_1, f(a_1))$ and $(a_2, f(a_2))$ on the curve has rise over run given by $\frac{f(a_2)-f(a_1)}{a_2-a_1}$, which, in this chapter, is being called the average rate of change of the function from $x = a_1$ to $x = a_2$.

So we have just deduced that a linear function f, that is a function with graph a straight line, has the property that the average rate of change of the function, no matter between which two values it is computed, is the same fixed value.

Let's go a step further and deduce from this a general formula for $f(x)$. We'll follow the idea of a previous practice exercise.

Suppose the ratio $\frac{f(a_2)-f(a_1)}{a_2-a_1}$ is always m for any two input values $x = a_1$ and $x = a_2$. Now let's choose two friendlier looking input values. Let's choose 0 (often a good choice in mathematics) and a general one, x. Then we have

$$\frac{f(x) - f(0)}{x - 0} = m.$$

Let b be the value of $f(0)$. (Most people call this value the y-intercept of the function.) Then our equation reads

$$\frac{f(x) - b}{x} = m,$$

which then yields $f(x) = mx + b$. This is the familiar equation of a line.

So we have just established that if f is a linear function, then the average rate of change of the function always computes to some constant value m and, moreover, f is given by $f(x) = mx + b$, with b the y-intercept of the function.

And to be logically tight, let's explore the converse.

Suppose a function f is given by $f(x) = mx + b$ for some constants m and b. Then for any two inputs $x = a_1$ and $x = a_2$, the average rate of change of the function is

$$\frac{f(a_2) - f(a_1)}{a_2 - a_1} = \frac{(ma_2 + b) - (ma_1 + b)}{a_2 - a_1}$$
$$= \frac{ma_2 - ma_1}{a_2 - a_1}$$
$$= m.$$

We see that the average rate of change of this function computes to the same constant value m for all pairs of input values.

In summary:

> A function f has constant average rate of change precisely if it is of the form $f(x) = mx + b$ for some constants m and b. The value m is the constant average rate of change (and b is the value of $f(0)$). Each such function has a straight line graph, and m is the slope of the line.

Comment. We have technically established that the equation of a line is indeed of the form $y = mx + b$. Our work above mimics the work usually presented in a beginning algebra class, which typically goes as follows.

Suppose a line has slope 7 and crosses the y-axis at 13. (See Figure 38.) Find an equation that must be true for a general point (x, y) to be on the line.

Well, for a point (x, y) to be on the line, the slope of the segment connecting $(0, 13)$ and (x, y) must be 7. Thus we require

$$\frac{y - 13}{x - 0} = 7$$

to be true. But there is a problem with this equation: we have subtly assumed that the point (x, y) is different from $(0, 13)$. The point $(0, 13)$ is itself on the line, but the equation $\frac{y - 13}{x} = 7$ is not valid for this point. But this is easily remedied by rewriting the equation as $y = 7x + 13$, which is a form of the equation that yields a true statement even for $x = 0$ and $y = 13$.

A valid equation of the line is thus $y = 7x + 13$.

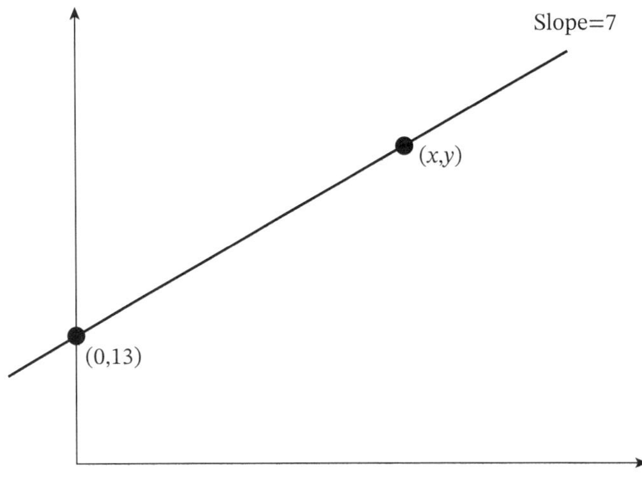

FIGURE 38

The following exercises are typical in a school curriculum.

Exercise. A cell phone company charges its customers a monthly flat fee of $30 and, in addition, $0.05 per minute of phone use that month. If t represents the number of minutes a customer uses one particular month, write down a formula that models $C(t)$, the amount charged to that customer that month.

Solution. We are told an initial value $C(0)$ and the value of a constant charge rate. We must have $C(t) = 0.05t + 30$. (This formula only approximates the call charge. It is not clear how fractional minutes of use are being handled, for example.) $\square$

Exercise. The same cell phone company has modified its charging rates. It still charges a monthly flat fee of $30 to each of its customers, but it charges $0.05 only for the first 300 minutes of phone use each month and then reduces the rate to $0.02 per minute for each minute of use thereafter. Describe a formula for $C(t)$, the amount charged to a customer after t minutes of use one month.

Solution. We still have $C(t) = 0.05t + 30$ for $0 \leq t \leq 300$, that is, for up to 300 minutes of use, but the formula changes to one of a different form for t greater than 300 minutes. It is a formula of the form $0.02(t-300)+b$ for some number b. (Notice how 300 is the "new zero" in this formula!) Now b must be the amount owed just before any fees are accrued for additional minutes of usage. We must have $b = C(300)$, which is $0.05 \times 300 + 30 = 45$.

So we have a split formula for $C(t)$:

$$C(t) = \begin{cases} 0.05t + 30 & \text{for } 0 \leq t \leq 300, \\ 0.02(t - 300) + 45 & \text{for } t > 300. \end{cases} \qquad \square$$

A function is called *piecewise linear* if its graph is composed of line segments and rays. Presenting such a function algebraically requires writing a linear function expression for a series of ranges of inputs (and it is usually annoying and tedious work to do this!)

MAA Featured Problem

(#20, AMC 12A, 2005)

For each x in $[0, 1]$, define

$$f(x) = \begin{cases} 2x & \text{if } 0 \leq x \leq \frac{1}{2}, \\ 2 - 2x & \text{if } \frac{1}{2} < x \leq 1. \end{cases}$$

Let $f^{[2]}(x) = f(f(x))$, and let $f^{[n+1]}(x) = f^{[n]}(f(x))$ for each integer $n \geq 2$. For how many values of x in $[0, 1]$ is $f^{[2005]}(x) = 1/2$?

A Personal Account of Solving This Problem

Curriculum Inspirations Strategy (`www.maa.org/ci`):

$$\boxed{\text{Strategy 2: Do something!}}$$

This question seems scary! For starters, the notation seems formidable. Can I unravel it a bit?

We are told $f^{[2]}(x) = f(f(x))$ is the two-fold composition of f with itself. Then we have:

$$f^{[3]}(x) = f^{[2]}(f(x)) = f(f(f(x))), \text{ the three-fold composition.}$$

$$f^{[4]}(x) = f^{[3]}(f(x)) = f(f(f(f(x)))), \text{ the four-fold composition.}$$

Oh! We have that $f^{[n]}$ is notation for the n-fold composition of f, namely, $f \circ f \circ \cdots \circ f$, n times.

Okay. Now the question wants us to play with the 2005th composition of f. Heavens!

The other part of the question that scares me is the description of the function itself (yet alone the 2005th iteration of it!) The only thing I can think to do right now is just to sketch its graph. See Figure 39.

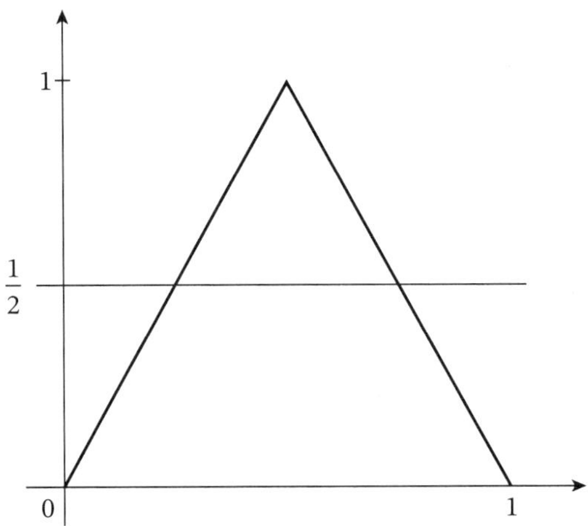

FIGURE 39

We are being asked to solve $f^{[2005]}(x) = \frac{1}{2}$. That's hard!

But I can solve $f(x) = \frac{1}{2}$. There are two solutions: $x = \frac{1}{4}$ and $x = \frac{3}{4}$. (I get this from working with $2x = \frac{1}{2}$ and $2 - 2x = \frac{1}{2}$. But we can also see these solutions from the graph.)

Can I start playing with compositions? Can I solve $f(f(x)) = \frac{1}{2}$?

Well, we just showed that if $f(\text{something}) = \frac{1}{2}$, then that something is $\frac{1}{4}$ or $\frac{3}{4}$. So $f(f(x)) = \frac{1}{2}$ gives $f(x) = \frac{1}{4}$ or $\frac{3}{4}$. We can see from the graph in Figure 40 that this has solutions $\frac{1}{8}$, $\frac{3}{8}$ and $\frac{5}{8}$, $\frac{7}{8}$.

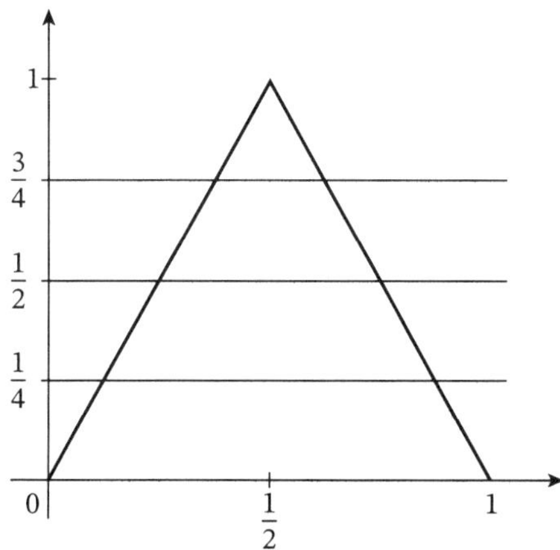

FIGURE 40

Ahh! I feel like I am on to something.

Solving $f(f(f(x))) = \frac{1}{2}$ means solving $f(f(\text{something})) = \frac{1}{2}$. That something must be $\frac{1}{8}$, $\frac{3}{8}$, $\frac{5}{8}$, or $\frac{7}{8}$; that is, we must have $f(x) = \frac{1}{8}$, $\frac{3}{8}$, $\frac{5}{8}$, or $\frac{7}{8}$. The graphs then show that x must be a sixteenth with odd numerator. We have found all the solutions to $f^{[3]}(x) = \frac{1}{2}$.

Carrying on, the equation $f^{[4]}(x) = \frac{1}{2}$ will have as solutions the fractions with denominator 32 and odd numerator, and so on, all the way up to $f^{[2005]}(x) = \frac{1}{2}$ with solutions the fractions with denominator 2^{2006} and odd numerators.

Was that the question?

No! The question wants less detail than this. It just wants the number of solutions to $f^{[2005]}(x) = \frac{1}{2}$. There are 2^{2006} fractions between 0 and 1 with denominator 2^{2006}, and half of them have odd numerators. There are thus a total of $\frac{1}{2} \cdot 2^{2006} = 2^{2005}$ solutions.

Additional Problems

32. (#5, AMC 12A, 2004) The graph of a line $y = mx + b$ is shown below.

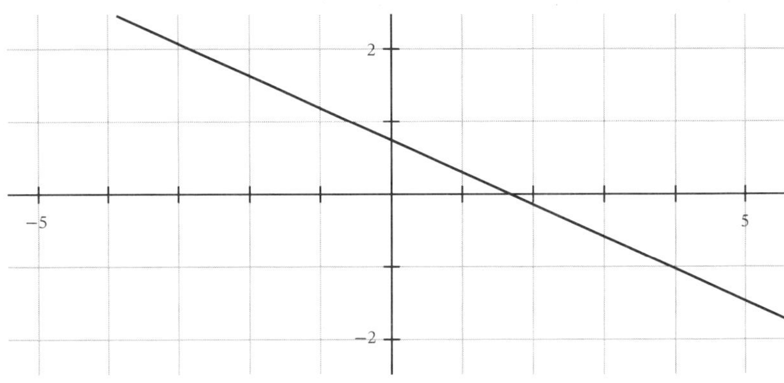

(This diagram is a redrawing of the original figure.)

Which of the following is true?
(A) $mb < -1$
(B) $-1 < mb < 0$
(C) $mb = 0$
(D) $0 < mb < 1$
(E) $mb > 1$

33. (#11, AMC 10B, 2003) A line with slope 3 intersects a line with slope 5 at the point $(10, 15)$. What is the distance between the x-intercepts of these two lines?
(A) 2
(B) 5
(C) 7
(D) 12
(E) 20

34. (#19, AMC 12A, 2002) The graph of the function f is shown below. How many solutions does the equation $f(f(x)) = 6$ have?
(A) 2
(B) 4
(C) 5
(D) 6
(E) 7

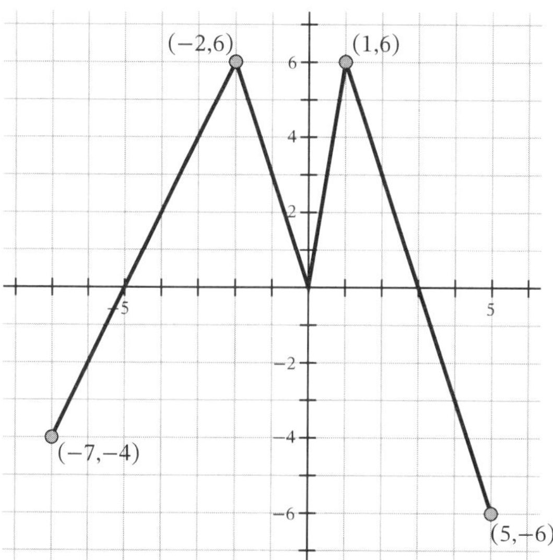

(This diagram is a redrawing of the original figure.)

35. (#24, AMC 12B, 2003) Positive integers a, b, and c are chosen so that $a < b < c$, and the system of equations

$$2x + y = 2003 \quad \text{and} \quad y = |x - a| + |x - b| + |x - c|$$

has exactly one solution. What is the minimum value of c?

(A) 668

(B) 669

(C) 1002

(D) 2003

(E) 2004

8

Quadratic Functions

An expression of the form $ax^2 + bx + c$ (or an algebraically equivalent form) with a, b, and c fixed numbers, $a \neq 0$, is called a *quadratic* expression in the variable x. (Thus $2x^2 + 7x - 8 = 9$ is an example of a quadratic equation and $f : \mathbb{R} \to \mathbb{R}$ given by $f(x) = 1 - x^2$ is an example of a quadratic function.)

The name "quadratic" is somewhat strange. The prefix *quad* means four and in what way is an expression of the form $ax^2 + bx + c$ connected to the number four? After all,

$$ax^3 + bx^2 + cx + d$$

is called a *cubic* expression because the highest power of x present is x cubed;

$$ax^4 + bx^3 + cx^2 + dx + e$$

is a *quartic* expression with x to the fourth power;

$$ax^5 + bx^4 + cx^3 + dx^2 + ex + f$$

is a *quintic* expression, and so on. (Is a seven-degree expression called "septic"?)

The answer is that scholars of ancient times realized that there is a deep geometric connection between quadratic equations and the symmetry of a square. A general method for solving a quadratic equation is to literally "make it square", a phrase which, in Latin, translates to *exquadrare*. These equations thus became known as the ones that can be solved by the square method.

Symmetry is a powerful tool in mathematics. For example, if I tell you that a rectangle has area 36 square units, you can tell me nothing more about that rectangle. Perhaps it is a 4×9 rectangle, or a $4\frac{1}{2} \times 8$

101

one. It is impossible to know. But if I then mention that the rectangle is actually symmetrical (that is, that it is a square), then you now know everything there is to know about that figure: it must be a 6 × 6 square.

The mathematical story of quadratics is really one about the power of symmetry. It is not typically presented in the curriculum this way. In this chapter we give a swift overview of both the algebra and graphing of quadratics from this perspective. Full details of this natural approach appear at `www.gdaymath.com/courses/`.

The Algebra of Quadratics

Some quadratic equations are straightforward to solve.

Practice. Solve $x^2 = 9$.

Reading the equation out loud, x squared is 9, suggests the geometry of the problem: a square of side length x has area 9. Clearly, that square has side length 3. We have the solution $x = 3$.

As negative side lengths are not permitted in geometry, this is the only geometric solution to the equation. But in the context of arithmetic, there is a second arithmetic value whose square is 9, namely, $x = -3$. So we actually have two numerical solutions: $x = 3$ or -3. See Figure 41.

Because geometric diagrams represent true arithmetic statements, we can be a little quirky and allow negative side lengths in our figures. This saves us the trouble of having to always distinguish between purely geometric and arithmetic solutions: we'll capture all arithmetic solutions through these generalized diagrams.

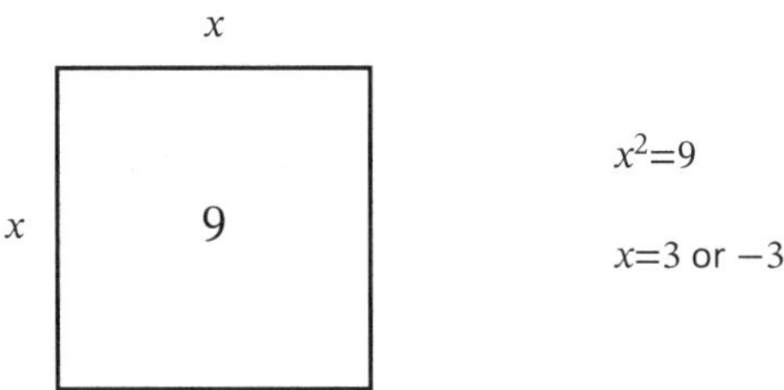

FIGURE 41

Practice. Solve $(x - 4)^2 = 25$.

This problem simply states that "something squared is 25". That something must be 5 or -5. Thus

$x - 4 = 5 \text{ or } -5$.

Adding 4 throughout gives

$x = 9 \text{ or } -1$.

Comment. It is tempting to write

$$(x - 4)^2 = 25,$$
$$x - 4 = \pm 5,$$
$$x = \pm 9.$$

But this is not correct. Consider avoiding the $\pm$ notation and taking an extra moment here and there to write out the word "or" to obviate this common error.

Practice. Solve $(2a + 1)^2 = 7$.

Although 7 is an awkward number, there is no conceptual difference in our approach to solving this problem.

$$(2a + 1)^2 = 7,$$
$$2a + 1 = \sqrt{7} \text{ or } -\sqrt{7},$$
$$2a = \sqrt{7} - 1 \text{ or } -\sqrt{7} - 1,$$

and so

$$a = \frac{\sqrt{7} - 1}{2} \text{ or } \frac{-\sqrt{7} - 1}{2}.$$

Now let's go up a notch in difficulty.

Practice. Solve $x^2 + 10x + 25 = 81$.

For those learning how to solve quadratic equations for the first time, this change of example is shocking! This particular quadratic equation seems to be of substantially different character to the previous examples. How might one approach solving it?

We do see in the equation an x^2 term. We have at least one square, an $x \times x$ square. But we have extra terms, $10x$ and 25, to tack onto this. Can we attach them to our $x \times x$ square and preserve the symmetry of our square? (We don't want to lose the power of symmetry!)

Can we adjoin "$10x$" in some symmetrical way?

Figure 42 shows that thinking of $10x$ as $5x + 5x$ is a good move, as we can now adjoin $5 \times x$ rectangles to the $x \times x$ in a symmetrical way.

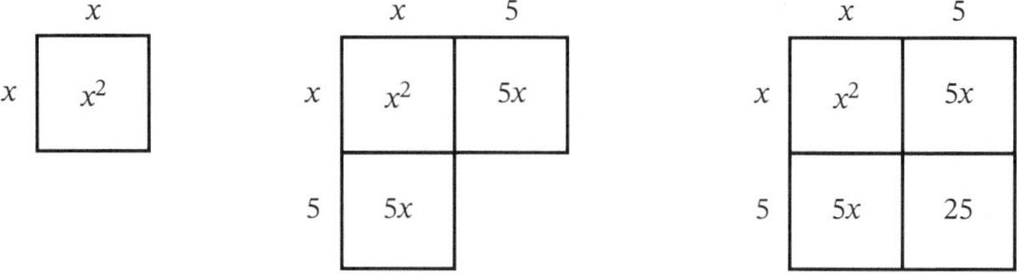

$$\text{F{\small IGURE}} \ 42$$

This gives an L-shaped figure, which looks like a large square with a corner missing. We see that corner is a 5×5 square of area 25, and we lucked out: the number 25 is the value we need to next contend with!

Figure 42 thus shows that $x^2 + 10x + 25$ is equivalent to an $(x+5) \times (x+5)$ square, and our practice problem is just the equation $(x+5)^2 = 81$ in disguise:

$$x^2 + 10x + 25 = 81,$$
$$(x + 5)^2 = 81,$$
$$x + 5 = 9 \text{ or } -9,$$
$$x = 4 \text{ or } -14.$$

Lovely!

Comment. In Figure 42 I've drawn all widths the same length and one might deduce that I wish to imply that x and 5 have the same value. That is not at all the case. As we do not know what the value of x shall be, it is immaterial what relative lengths we use in these diagrams. But do feel free to make your own diagrams look more extreme if you wish. Alternative Figure 42 shows an example with length 5 longer than length x. All the diagrams one draws are sure to illustrate the same underlying symmetry.

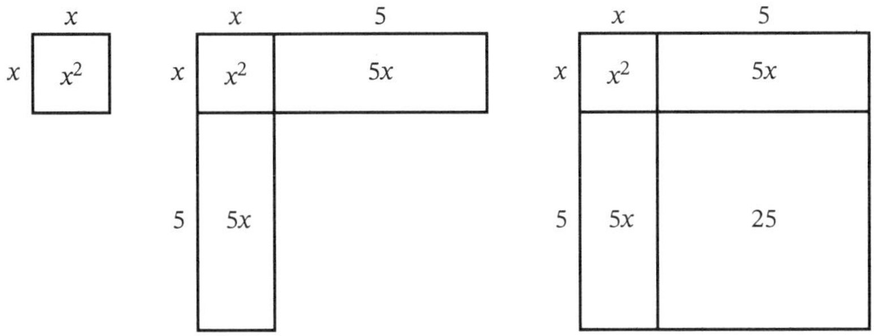

Alternative Figure 42

Practice. Solve $x^2 - 16x + 64 = 1$.

Think of this as an x^2 piece, two $-8x$ pieces, and an extra piece that, hopefully, matches the number 64. Figure 43 shows all is good!

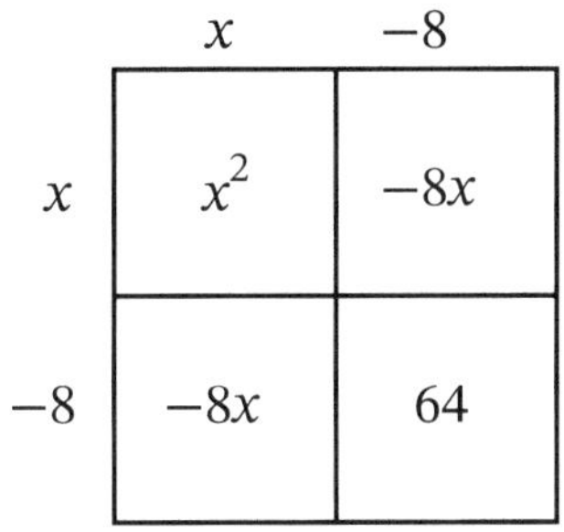

Figure 43

$$x^2 - 16x + 64 = 1,$$
$$(x - 8)^2 = 1,$$
$$x - 8 = 1 \text{ or } -1,$$
$$x = 9 \text{ or } 7.$$

This next problem has a problem.

Practice. Solve $x^2 - 6x + 7 = 23$.

Drawing the square with pieces x^2 and $-3x$ and $-3x$ (Figure 44) shows we have a mismatch with the final corner of the square. The geometry wants the number 9, and we have the number 7. Hmm.

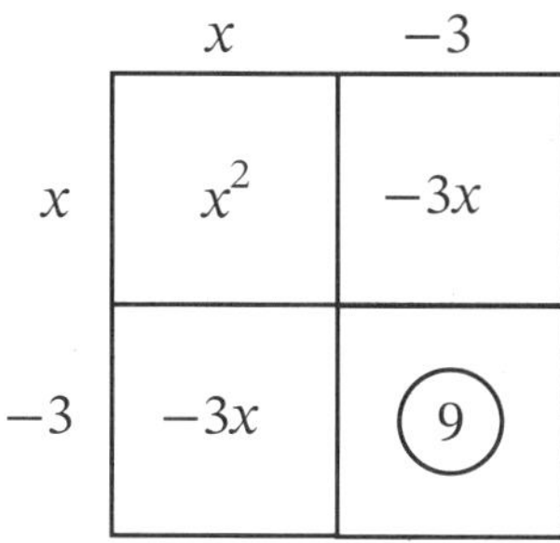

FIGURE 44

Well, if the problem wants the number 9, let's make it appear! Let's add 2 to the 7 on the left. And to keep the algebra balanced this means we need to add 2 to the 23 on the right as well.

$$x^2 - 6x + 7 = 23,$$
$$x^2 - 6x + 9 = 25,$$
$$(x - 3)^2 = 25,$$
$$x - 3 = 5 \text{ or } -5,$$
$$x = 8 \text{ or } -2.$$

Beautiful!

Carrying on with this swift overview, let's go up yet another notch of difficulty.

Practice. Solve $x^2 + 5x + 1 = 7$.

This problem has us draw a square with pieces x^2 and $\frac{5}{2}x$ and $\frac{5}{2}x$. The 1 might be a mismatch of number, but from the previous example, we know how to handle that hiccup. This approach, of course, works (see Figure 45), but it is not fun working with fractions. Is there a way to avoid fractions?

The trouble occurred with the middle term of the quadratic, the $5x$, having an odd coefficient. Can we avoid an odd middle coefficient?

$$x^2+5x+1=7$$

$$x^2+5x+\frac{25}{4}=7+\frac{21}{4}$$

$$\left(x+\frac{5}{2}\right)^2=12\frac{1}{4}$$

$$\vdots$$

FIGURE 45

Yes! Let's multiply the equation through by 2 and solve instead

$$2x^2 + 10x + 2 = 14.$$

The first piece is now $2x^2$ which comes from a $\sqrt{2}x \times \sqrt{2}x$ square. (We don't want to work with an $x \times 2x$ piece as that is no longer square and it will ruin our symmetrical square technique.)

Figure 46 shows that we can push matters and make this version of the equation work, but dealing with irrational numbers is even more awkward than working with just fractions.

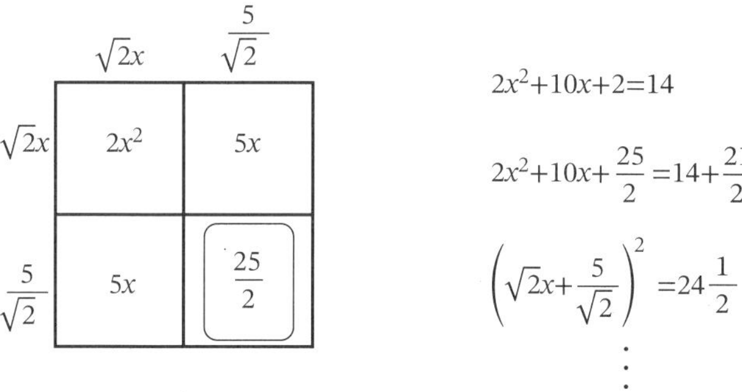

$$2x^2+10x+2=14$$

$$2x^2+10x+\frac{25}{2}=14+\frac{21}{2}$$

$$\left(\sqrt{2}x+\frac{5}{\sqrt{2}}\right)^2=24\frac{1}{2}$$

$$\vdots$$

FIGURE 46

The problem here is that the 2 in $2x^2$ is not a perfect square. Going back to $x^2+5x+1=7$, is there a way to modify the equation so that the coefficient of the middle term is even *and* the coefficient of the leading term is a perfect square?

A flash of insight suggests to multiply through by 4. It leads to a square figure with integer components. (See Figure 47.)

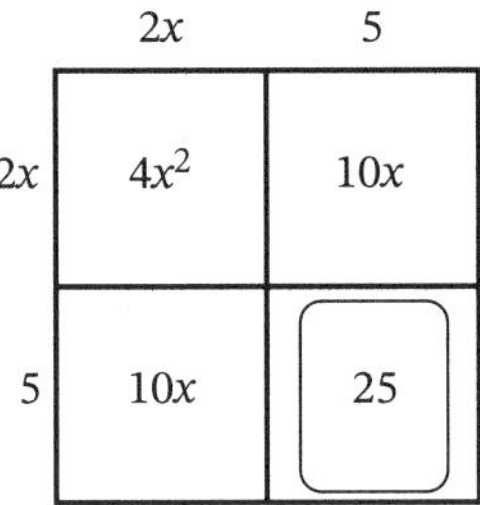

FIGURE 47

We have

$$x^2 + 5x + 1 = 7,$$
$$4x^2 + 20x + 4 = 28,$$
$$4x^2 + 20x + 25 = 49,$$
$$(2x + 5)^2 = 49,$$
$$2x + 5 = 7 \text{ or } -7,$$
$$2x = 2 \text{ or } -12,$$
$$x = 1 \text{ or } -6.$$

Great!

Now here's an additional hiccup.

Practice. Solve $3x^2 - 7x + 1 = 11$.

This example has two challenges: the leading coefficient, 3, is not a perfect square and the coefficient of the middle term, -7, is odd. We can push the square method through if we are willing to work with irrational numbers and fractions. But what could we do to avoid them?

Multiplying through by 3 is one way to fix the first issue:

$$9x^2 - 21x + 3 = 33.$$

But the coefficient of the middle term is still odd. We can fix this by multiplying through by 4, and doing so does not ruin the leading coefficient being a perfect square:

$$36x^2 - 84x + 12 = 132.$$

The numbers are surprisingly large but are perfectly friendly for the square method. (See Figure 48.)

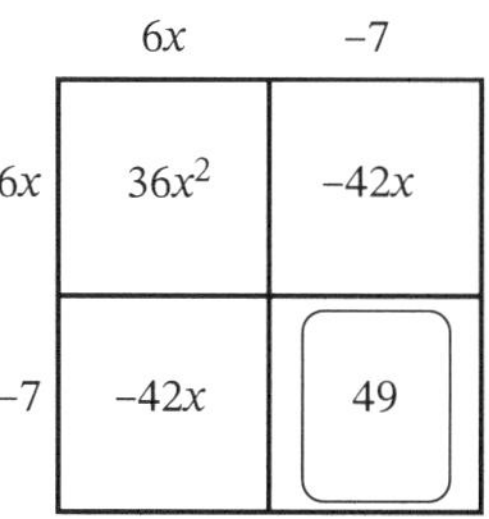

FIGURE 48

$$3x^2 - 7x + 1 = 11,$$
$$9x^2 - 21x + 3 = 33,$$
$$36x^2 - 84x + 12 = 132,$$
$$36x^2 - 84x + 49 = 169,$$
$$(6x - 7)^2 = 169,$$
$$6x - 7 = 13 \text{ or } -13,$$
$$6x = 20 \text{ or } -6,$$
$$x = \frac{10}{3} \text{ or } -1.$$

In summary:

> To solve $ax^2 + bx + c = d$ perhaps multiply through by a and through by 4 to avoid irrational terms and fractions. The method of completing a square applied to the resulting equation then gives the solutions to the quadratic equation.

Comment. Most school curricula have students first rewrite quadratic equations as a quadratic expression set equal to 0. As we have seen, there is no need to do this. But if you do, our approach then yields the famous quadratic formula students are often required to memorize.

To solve $ax^2 + bx + c = 0$ first multiply through by a and then by 4 and solve instead $4a^2x^2 + 4abx + 4ac = 0$. This gives the square shown in Figure 49.

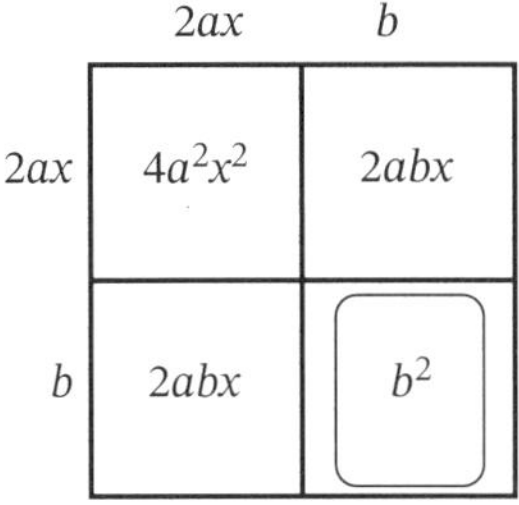

FIGURE 49

Now read through the following steps slowly:

$$ax^2 + bx + c = 0,$$

$$4a^2x^2 + 4abx + 4ac = 0,$$

$$4a^2x^2 + 4abx + b^2 = b^2 - 4ac,$$

$$(2ax + b)^2 = b^2 - 4ac,$$

$$2ax + b = \sqrt{b^2 - 4ac} \text{ or } -\sqrt{b^2 - 4ac},$$

$$2ax = -b + \sqrt{b^2 - 4ac} \text{ or } -b - \sqrt{b^2 - 4ac},$$

$$x = \frac{-b + \sqrt{b^2 - 4ac}}{2a} \text{ or } \frac{-b - \sqrt{b^2 - 4ac}}{2a}.$$

We have the quadratic formula.

Of course, there is no need to memorize this formula. Just draw a square! (I do understand, however, that the quadratic formula is deemed such a key element of many algebra curricula that it is usually expected that students will know it by heart.)

Comment. We have shown that every quadratic equation is equivalent to an equation of the form "something squared equals a number". If that number is positive, then the quadratic equation has two roots. This has been the case for every example in this chapter. If the number is 0, then the quadratic equation has only one solution: the number 0 has only one square root. And if that number is negative, then that number has no square roots and so the equation has no (real) solution.

Exercise. Solve $x^2 + 2x + 2 = 1$.

Solution. Drawing a square shows

$$x^2 + 2x + 2 = 1,$$
$$x^2 + 2x + 1 = 0,$$
$$(x + 1)^2 = 0,$$
$$x + 1 = 0,$$
$$x = -1.$$

The equation has precisely one solution. $\square$

Exercise. Solve $x^2 - 4x + 7 = 0$.

Solution. Drawing a square shows

$$x^2 - 4x + 7 = 0,$$
$$x^2 - 4x + 4 = -3,$$
$$(x - 2)^2 = -3.$$

This quadratic equation has no real solution. $\square$

Final comment. Multiplying a quadratic expression through by 4 just once might not always be quite enough to obviate fractions. (For example, try solving $4x^2 + x + 1 = 3$.) One can always multiply through by 4 an additional time.

Graphing Quadratics

The graph of the basic quadratic equation $y = x^2$ is a symmetrical U-shaped curve. See Figure 50.

Comment. Many curves in mathematics look "U-shaped". This descriptor is being used in a very loose sense. The precise shape of the U-shaped curve for this quadratic graph turns out to match one of the curves studied in classic Greek geometry, namely, the *parabola*. We'll discuss this at the end of this chapter.

Our study of the transformation of graphs shows that the graphs of $y = (x - k)^2$ and $y = x^2 + k$, for a given number k, are just translations of this graph and so they too have symmetrical U-shaped curves for their graphs.

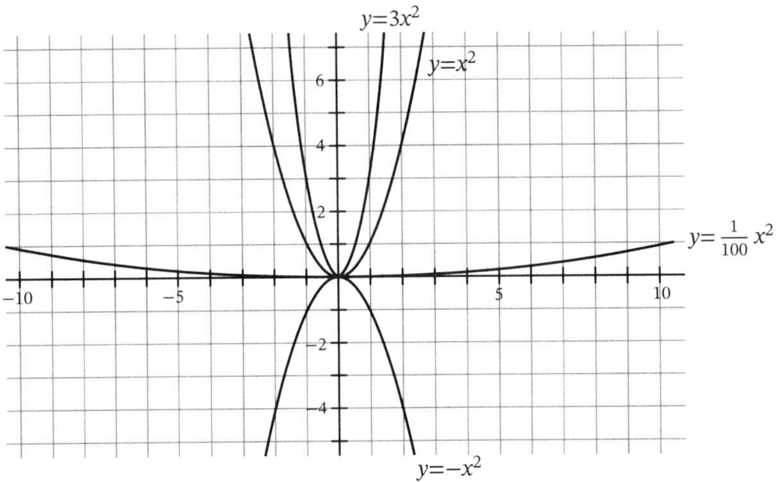

FIGURE 50

The graph of $y = 3x^2$ with all y-coordinate values tripled is a steep U-shaped curve, the graph of $y = \dfrac{1}{100}x^2$ is a shallow U-shaped curve, and the graph of $y = -x^2$ is an inverted U. In each case, the graph remains symmetrical in shape.

In summary:

> The graph of $y = ax^2$ is a symmetrical U-shaped graph with a a measure of the steepness of the curve. (If a is negative, the U-shaped curve is inverted.) The *vertex* of the quadratic (the base of the U) is at the origin.

The graph of $y = a(x-k)^2 + h$ is a translate of the graph of $y = ax^2$. Here $x = k$ is behaving like 0 for the x-values and all y-coordinates are increased by h in value. (Thus the vertex of the quadratic is at position (k, h).)

Exercise. Is "U" the correct description of the graph of $y = x^2$? Do the arms of the graph of $y = x^2$ ever become vertical?

Solution. If the right arm of the graph was vertical say at $x = N$, then where is the point for $x = N + 1$ on the graph? The arms can never be vertical. (But the loose language of "U-shaped" still evokes a helpful image.)   □

Exercise. The U-shaped graph of $y = x^2$ is rotated clockwise about the origin through an angle of $0.001°$. Does the left arm of the rotated curve cross the y-axis?

Hint towards solution. Instead, think of drawing a copy of the y-axis rotated about the origin counterclockwise though an angle of $0.001°$. Must this nonvertical line intercept the left arm of the $y = x^2$ graph?

Here's a lovely feature of quadratics:

> Every quadratic equation $y = ax^2 + bx + c$ can be rewritten in the form $y = a(x - k)^2 + h$ for some numbers k and h, and so the graph of every quadratic equation is a symmetrical U-shaped curve with vertex at some position in the plane, namely, (k, h).

This is actually a stunning result. For instance, the equation $y = x^2$ has a symmetrical U-shaped curve for its graph, and the equation $y = x$ has a diagonal line for its graph. Now consider the equation $y = x^2 + x$. Each point in the graph of this equation, $(x, x^2 + x)$, is a point of the U-shaped graph of $y = x^2$ shifted upward if x is positive and shifted downward if x is negative. This is asymmetrical. How can the shape of the graph of $y = x^2 + x$ possibly be another symmetrical U-shaped graph? See Figure 51.

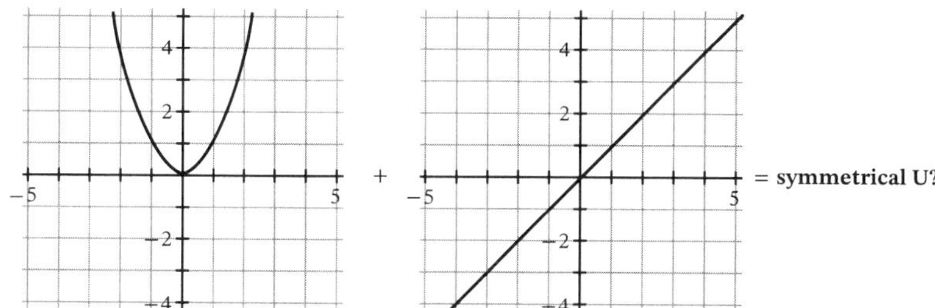

FIGURE 51

Exercise. Plot some points of the graph of $y = x^2 + x$ by hand. See that its graph does indeed seem to be another symmetrical U-shaped graph. Can you identify the vertex of the graph?

Really do try this!

The algebraic work of the previous section establishes our general claim. We'll use the square method.

Example. Consider the equation $y = x^2 - 4x + 3$.

Our square method suggests rewriting this as $y + 1 = x^2 - 4x + 4$, that is, as $y + 1 = (x - 2)^2$. So we have $y = (x - 2)^2 - 1$. Thus the graph of $y = x^2 - 4x + 3$ is the graph of $y = x^2$ translated so that its vertex lies at $x = 2$, $y = -1$.

Example. Consider the equation $y = 3x^2 + x - 7$. Let's rewrite it in the form $y = a(x - k)^2 + h$ for some values k and h. (This is a deliberately awkward example designed to show how the process can be conducted in general.)

Multiplying through by 3 and then by 4 gives $12y = 36x^2 + 12x - 84$. The square method shows this is equivalent to $12y + 85 = (6x + 1)^2$. This is equivalent to

$$12y + 85 = (6x + 1)(6x + 1) = 36\left(x + \frac{1}{6}\right)^2,$$

which can be rewritten

$$y = 3\left(x + \frac{1}{6}\right)^2 - \frac{85}{12}.$$

Thus the graph of $y = 3x^2 + x - 7$ is just a translation of the steep U-shaped graph of $y = 3x^2$.

Exercise. Rewrite $y = x^2 + x$ in a form that shows that its graph is indeed a translate of the symmetrical U-shaped graph of $y = x^2$.

Solution. Follow the square method to get $y = \left(x + \frac{1}{2}\right)^2 - \frac{1}{4}$. $\square$

Exercise. If you are feeling game, use the square method to rewrite $y = ax^2 + bx + c$ in the form $y = a(x - k)^2 + h$ for some values k and h.

Solution. After some hefty algebra, one obtains $y = a\left(x + \frac{b}{2a}\right)^2 + c - \frac{b^2}{4a^2}$. $\square$

Having done this algebraic work two or three time (or if we're gung-ho, more than three times), my advice is to never do it again! Performing this algebraic work on algebraic equations is an onerous approach to exploring the graphs of quadratics. All we need to observe is the following:

> The graph of any quadratic equation $y = ax^2 + bx + c$ is a symmetrical U-shaped graph.

The stand-out feature here is symmetry. Symmetry is a powerful mathematical friend.

Practice. Sketch a graph of $y = (x - 3)(x - 7) + 10$.

If we were to expand the brackets, we'd see that this equation is equivalent to a quadratic equation $y = ax^2 + bx + c$ and so it has a graph that is a symmetrical U-shaped curve.

Looking at the equation, we can't help but think that $x = 3$ and $x = 7$ are interesting x-values. In fact, they each give the value $y = 10$. Thus $(3, 10)$ and $(7, 10)$ are two symmetrical points on a symmetrical U-shaped curve. The line of symmetry of the curve must be at $x = 5$, halfway between these two interesting x-values. This also means that the vertex of the U is at $x = 5$.

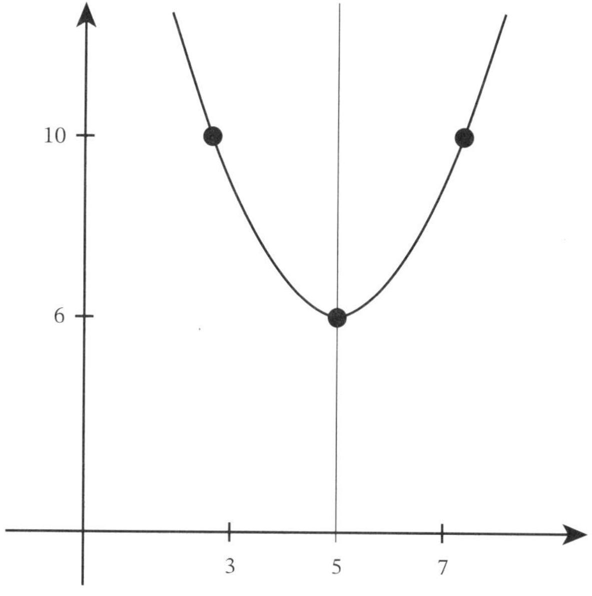

FIGURE 52

Do we know the y-coordinate of the vertex? Well, let's just put $x = 5$ into the equation! This gives $y = (2)(-2) + 10 = 6$. The vertex of the U is at $(5, 6)$. This is enough information to give a good quick sketch of the graph. (See Figure 52.)

Practice. Sketch the graph of $y = x^2 - 12x + 22$.

The previous practice example had two interesting x-values directly in sight. No such values seem to be immediate in this case. Let's work with the x part of the equation, the $x^2 - 12x$ component, and ignore the $+22$ for now. Let's see if we can make any interesting x-values naturally appear.

I can only think to rewrite $x^2 - 12x$ as $x(x - 12)$. Ahh! Our equation can be rewritten

$$y = x(x - 12) + 22$$

and we see that $x = 0$ and $x = 12$ are both interesting; they both give $y = 22$, and so two symmetrical points on the symmetrical U-shape graph. Thus the line of symmetry is at $x = 6$ and the vertex of the U has x-coordinate $x = 6$ and matching y-coordinate $y = 6(-6) + 22 = -14$. This is enough information to sketch the curve! (Figure 53 portrays all the correct information, except for a correct sense of scale. Is that important for our current work?)

One can always provide extra details about the graph if one wishes. For example, to find where this graph crosses the x-axis set $y = 0$ in the equation $y = x^2 - 12x + 22$. The square method then gives

$$x^2 - 12x + 22 = 0,$$

$$x^2 - 12x + 36 = 14,$$

$$(x - 6)^2 = 14,$$

and so $x = 6 + \sqrt{14}$ and $x = 6 - \sqrt{14}$ are its x-intercepts. (Is this extra information illuminating?)

Practice. Sketch $y = 5 + 10x - 8x^2$.

To see interesting x-values rewrite the equation as

$$y = 2x(-4x + 5) + 5$$

(or as $y = x(10 - 8x) + 5$, so some other equally illuminating form) to see $x = 0$ and $x = \frac{5}{4}$ both give $y = 5$. We have two symmetrical points on

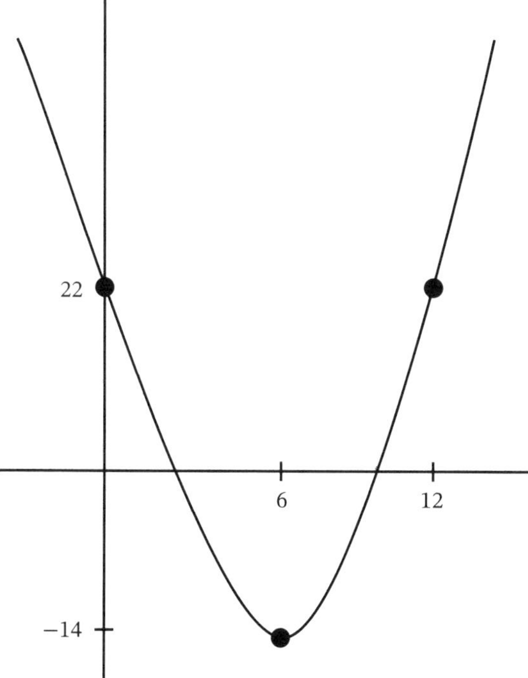

FIGURE 53

a symmetrical U-shaped curve. The line of symmetry must be halfway between them, at $x = \frac{5}{8}$. Here $y = 5 + 10 \cdot \frac{5}{8} - 8 \cdot \frac{25}{64} = 8\frac{1}{8}$.

Notice that the vertex has height $8\frac{1}{8}$, a value larger than the heights of our two symmetrical points at height 5. We must have an upside-down U-shaped curve. (And this makes sense since the steepness of this quadratic is -8, a negative number.)

We're all set to sketch the graph of the equation. (See Figure 54.)

Practice. Sketch $y = 33\frac{1}{3}\left(x + 7\frac{1}{2}\right)^2 - 100$.

Let's not forget we can see through some equations right away. The graph of this equation matches the graph of $y = 33\frac{1}{3}x^2$, a very steep upward-facing U-shaped curve. It is translated so that $x = -7\frac{1}{2}$ behaves

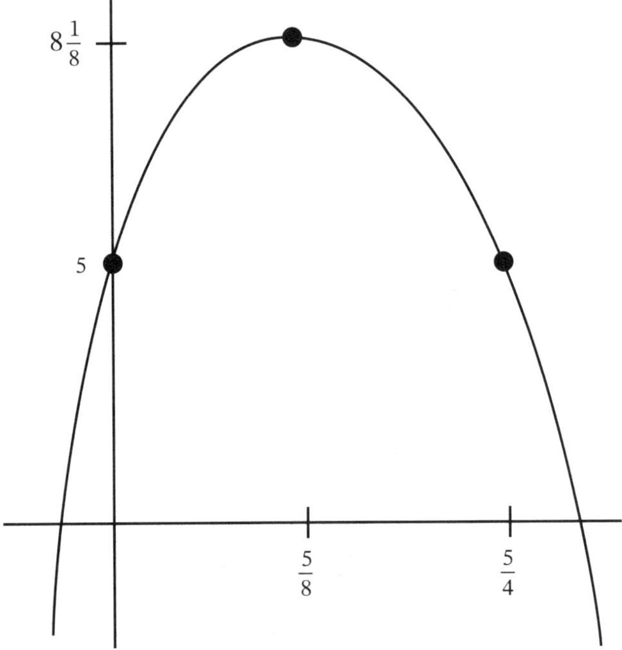

FIGURE 54

like 0 and is also translated 100 units downwards. The vertex of this curve is thus now at $\left(-7\frac{1}{2}, -100\right)$.

Exercise. What is the line of symmetry of the graph of $y = ax^2 + bx + c$?

Solution. We have $y = x(ax + b) + c$ and we see that $x = 0$ and $x = -\frac{b}{a}$ give symmetrical points on the symmetrical graph (they both give $y = c$). The line of symmetry of the curve must be halfway between them, which is at $x = \frac{1}{2} \cdot \left(-\frac{b}{a}\right) = -\frac{b}{2a}$. (This is a formula students are often expected to memorize.) $\qquad\square$

Exercise. Find a value k so that the graph of $y = 2x^2 - 18x + k$ just touches the x-axis.

Solution. We have $y = 2x(x - 9) + k$ and so the line of symmetry of the graph must be at $x = \frac{9}{2}$. This is where the vertex of the graph occurs.

For the graph to just touch the x-axis we must have this vertex on the axis; that is, we must have $y = 0$ at the vertex.

Now at $x = \frac{9}{2}$ we get $y = 2 \cdot \frac{9}{2}\left(\frac{9}{2} - 9\right) + k = -\frac{81}{2} + k$. This shows we need to set $k = \frac{81}{2}$. $\qquad\square$

There is no need to memorize any formulas when graphing quadratic equations: just look for interesting x-values that give symmetrical points on the graph and all the rest follows by just using your basic wits.

Exercise. Find the equation for a quadratic function f that has value 0 at $x = -5$ and $x = 11$ and has 200 as its largest output.

Solution. We want a quadratic formula that gives 0 at $x = -5$ and $x = 11$. Certainly $f(x) = (x + 5)(x - 11)$ has this property. But the graph of f is meant to have largest value 200. So we must have an upside-down U shaped graph: our formula yields an upward-facing U-curve. Also, we've ignored a possible steepness factor.

So let's work with $f(x) = a(x + 5)(x - 11)$ for some (negative) value a.

The maximum output of this function occurs halfway between $x = -5$ and $x = 11$, that is, at $x = 3$. So we need $200 = a(3+5)(3-11)$; that is, we need $a = -\frac{200}{64} = -\frac{25}{8}$. Thus f defined by $f(x) = -\frac{25}{8}(x + 5)(x - 11)$ defines a quadratic function with the desired properties. $\qquad\square$

This exercise shows that if a quadratic graph has x-intercepts $x = p$ and $x = q$, then it must be given by an equation of the form $y = a(x - p)(x - q)$ for some value a.

Comment. If $x^2 + bx + c$ has roots r_1 and r_2 (that is, these are the values of x that give $x^2 + bx + c = 0$), then we must have

$$x^2 + bx + c = (x - r_1)(x - r_2).$$

Expanding shows that

$$x^2 + bx + c = x^2 - (r_1 + r_2)x + r_1 r_2.$$

Thus we conclude that $c = r_1 r_2$ and $b = -(r_1 + r_2)$. (These two conclusions follow by trying $x = 0$ and then $x = 1$ in the equation $x^2 + bx + c = x^2 - (r_1 + r_2)x + r_1 r_2$.)

Many textbook and competition problems about quadratics make use of this relationship between the roots of a quadratic expression and its coefficients.

Exercise. If $ax^2 + bx + c$ has roots r_1 and r_2, what now is the relationship between the roots and the coefficients?

Solution. Look at $ax^2 + bx + c = a(x - r_1)(x - r_2)$. For $x = 0$ we see $c = ar_1r_2$ and for $x = 1$ we see

$$a + b + c = a(1 - r_1)(1 - r_2),$$
$$a + b + c = a - a(r_1 + r_2) + ar_1r_2,$$
$$b = -a(r_1 + r_2).$$

So $r_1r_2 = \dfrac{c}{a}$ and $r_1 + r_2 = -\dfrac{b}{a}$. $\qquad\qquad\square$

Final comment. It is not at all obvious that the shape of any quadratic graph has the property of being a *parabola*, the shape of one of the curves that arises from slicing a cone in classic Greek geometry. This turns out to be true, but it requires proof.

A parabola is also described as the set of points in the plane equidistant from a given line L, called the *directrix* of the parabola, and a given point F, called its *focus*. (See Figure 55.)

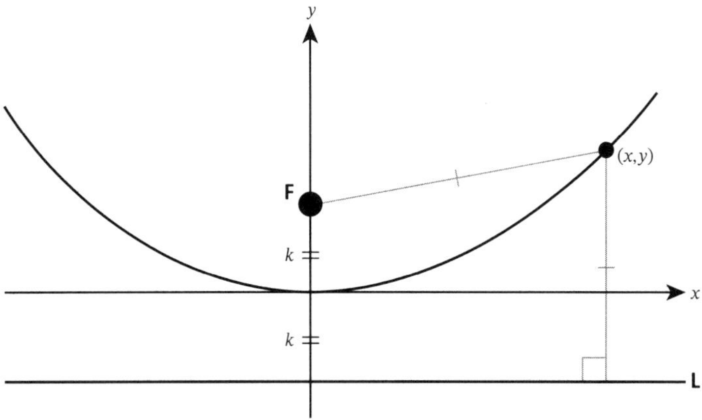

FIGURE 55

It is not obvious that each curve constructed this way also matches a particular planar slice of a cone, as ancient Greek scholars observed. (It can be proved so.)

What we can do now is prove that any quadratic graph has a natural focus and directrix associated with it and so satisfies the equidistance description of a parabola.

Through vertical and horizontal translations (and maybe a reflection) we can assume that our quadratic curve is a graph of an equation of the form $y = ax^2$ for some nonzero number a.

By symmetry, we might guess that a possible directrix L for the curve is a horizontal line and a possible focus F is a point on the vertical axis. Also, since the origin, the vertex of the curve, is meant to be equidistant from F and L, we might also guess that the line L is k units below the x-axis (and so has equation $y = -k$) and F is k units above it (and so has coordinates $(0, k)$) for some value k.

Let's go with these guesses and see if there is a possible candidate value for k.

Let $P = (x, y)$ be a point equidistant from F and L. Then, looking at Figure 55, we see

$$\sqrt{x^2 + (y - k)^2} = y + k.$$

This gives

$$x^2 + (y - k)^2 = (y + k)^2$$

from which we obtain $y = \dfrac{1}{4k}x^2$. This matches the equation $y = ax^2$ if we choose $k = \dfrac{1}{4a}$.

All right. So set L to be the line $y = -\dfrac{1}{4a}$ and F to be the point $\left(0, \dfrac{1}{4a}\right)$.

One now checks two things. (Each is essentially a repeat of the algebra we just conducted.)

If $P = (x, y)$ is a point with coordinates satisfying $y = ax^2$, then P in indeed equidistant from L and F.

If $Q = (x, y)$ is a point equidistant from L and F, then its coordinates satisfy the equation $y = ax^2$.

Thus the graph of $y = ax^2$ matches precisely the set of points equidistant from a focus and a directrix. Quadratic graphs are indeed parabolas.

MAA Featured Problem

(#25, AMC 12B, 2002)

Let $f(x) = x^2 + 6x + 1$, and let R denote the set of points (x, y) in the coordinate plane such that

$$f(x) + f(y) \le 0 \qquad \text{and} \qquad f(x) - f(y) \le 0.$$

Which of the following is closest to the area of R?

    (A) 21

    (B) 22

    (C) 23

    (D) 24

    (E) 25

A Personal Account of Solving This Problem

Curriculum Inspirations Strategy (`www.maa.org/ci`):

> Strategy 1: Engage in successful flailing.

This question seems strange. I see a quadratic function. And some quick scratch-work shows that it is an upward-facing U-shaped graph with vertex at $x = -3$. (In fact, the vertex is at $(-3, -8)$.) But I am not sure if this is helpful.

It seems I am meant to look at two values on this graph, an $f(x)$ value and an $f(y)$ value. Should I draw two copies of the same graph? That feels confusing and I am not even clear what I would do with two copies.

Can I instead play with the two inequalities?

Adding them (adding two nonpositive numbers) is sure to give a nonpositive result. So we conclude $2f(x) \le 0$; that is, $f(x) \le 0$.

Now $f(x) = 0$ when $x^2 + 6x + 1 = 0$, that is, when $(x + 3)^2 = 8$. So $f(x) \le 0$ for $-3 - \sqrt{8} \le x \le -3 + \sqrt{8}$. They are decidedly unfriendly numbers!

Hmm. What else can I do?

Well, $f(x) + f(y) \leq 0$ reads

$$x^2 + 6x + 1 + y^2 + 6y + 1 \leq 0.$$

Is this recognizable?

Well, if we rewrite it as $(x+3)^2 + (y+3)^2 \leq 8 + 8 = 16$, then we see $f(x) + f(y) \leq 0$ represents the set of all points (x, y) in the interior of a circle of radius $\sqrt{16} = 4$ with center $(-3, -3)$. They are nice numbers! Could the question have been constructed to be nice?

Okay then, what about $f(x) - f(y) \leq 0$, that is, $f(x) \leq f(y)$? This reads

$$x^2 + 6x + 1 \leq y^2 + 6y + 1$$

which is equivalent to

$$(x+3)^2 \leq (y+3)^2.$$

Okay, we want to examine the area of the region R defined by all points (x, y) satisfying

$$(x+3)^2 + (y+3)^2 \leq 16 \quad \text{and} \quad (x+3)^2 \leq (y+3)^2.$$

Can we just say that $x = -3$ and $y = -3$ are each behaving like 0 and just work with points (x, y) satisfying

$$x^2 + y^2 \leq 16 \quad \text{and} \quad x^2 \leq y^2$$

instead? (This is just a translation of R, three units in the x-direction and three units in the y-direction, and so its area is unchanged.)

Okay, I recognize $x^2 + y^2 \leq 16$ as the interior of a circle of radius 4. What is the region defined by $x^2 \leq y^2$?

The boundary of this region is given by $x^2 = y^2$. This means either $x = y$ or $x = -y$. Ahh! These are the diagonal lines $y = x$ and $y = -x$ through the origin. So the set of points satisfying (x, y) that satisfy $x^2 \leq y^2$ lie in the triangular wedges between these lines.

Which of the four wedges?

Well, when $x = 0$ the inequality $x^2 \leq y^2$ will hold. So we must be working with the two wedges that contain the y-axis. So the region we need to consider is two quarters of a circle of radius 4. See Figure 56.

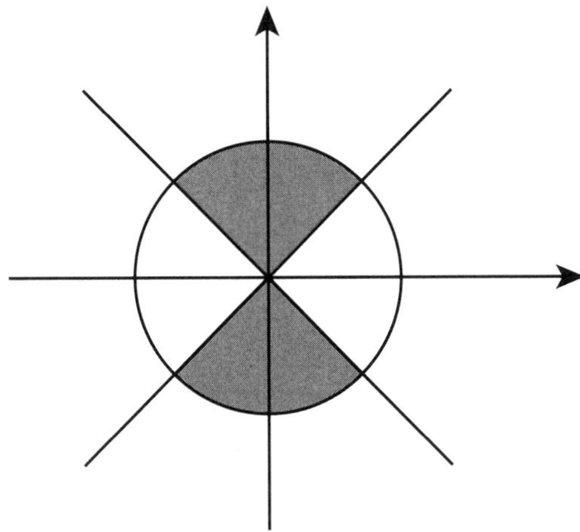

FIGURE 56

The area of region R is thus $\frac{1}{2} \cdot \pi 4^2 = 8\pi \approx 8 \times 3.1 = 24.8$. The answer is option (E). Wow!

Additional Problems

36. (#10, AMC 10A, 2002) What is the sum of the all the roots of $(2x + 3)(x - 4) + (2x + 3)(x - 6) = 0$?

  (A) $\frac{7}{2}$

  (B) 4

  (C) 5

  (D) 7

  (E) 13

37. (#10, AMC 10B, 2002) Suppose that a and b are nonzero real numbers and that the equation $x^2 + ax + b = 0$ has solutions a and b. What is the pair (a, b)?

  (A) $(-2, 1)$

  (B) $(-1, 2)$

  (C) $(1, -2)$

  (D) $(2, -1)$

  (E) $(4, 4)$

38. (#5, AMC 10A, 2003) Let d and e denote the solutions of $2x^2 + 3x - 5 = 0$. What is the value of $(d-1)(e-1)$?

   (A) $-\dfrac{5}{2}$

   (B) 0

   (C) 3

   (D) 5

   (E) 6

39. (#10, AMC 10A, 2005) There are two values of a for which the equation $4x^2 + ax + 8x + 9 = 0$ has only one solution for x. What is the sum of those values of a?

   (A) -16
   (B) -8
   (C) 0
   (D) 8
   (E) 20

40. (#16, AMC 10B, 2005) The quadratic equation $x^2 + mx + n = 0$ has roots that are twice those of $x^2 + px + m = 0$, and none of m, n, and p is 0. What is the value of n/p?

   (A) 1
   (B) 2
   (C) 4
   (D) 8
   (E) 16

41. (#14, AMC 10B, 2006) Let a and b be the roots of the equation $x^2 - mx + 2 = 0$. Suppose that $a + (1/b)$ and $b + (1/a)$ are the roots of the equation $x^2 - px + q = 0$. What is q?

   (A) $\dfrac{5}{2}$

   (B) $\dfrac{7}{2}$

   (C) 4

   (D) $\dfrac{9}{2}$

   (E) 8

42. (#18, AMC 10A, 2003) What is the sum of the reciprocals of the roots of the equation

$$\frac{2003}{2004}x + 1 + \frac{1}{x} = 0?$$

(A) $-\dfrac{2004}{2003}$

(B) -1

(C) $\dfrac{2003}{2004}$

(D) 1

(E) $\dfrac{2004}{2003}$

43. (#12, AMC 12A, 2002) Both roots of the quadratic equation

$$x^2 - 63x + k = 0$$

are prime numbers. How many values of k are possible?

(A) 0

(B) 1

(C) 2

(D) 4

(E) more than 4

44. (#8, AMC 12B, 2005) For how many values of a is it true that the line $y = x + a$ passes through the vertex of the parabola $y = x^2 + a^2$?

(A) 0

(B) 1

(C) 2

(D) 10

(E) infinitely many

45. (#12, AMC 12B, 2006) The parabola $y = ax^2 + bx + c$ has vertex (p, p) and y-intercept $(0, -p)$ where $p \neq 0$. What is b?

(A) $-p$

(B) 0

(C) 2

(D) 4

(E) p

46. (#13, AMC 12, 2001) The parabola with equation $y = ax^2 + bx + c$ and vertex (h, k) is reflected about the line $y = k$. This results in the parabola with equation $y = dx^2 + ex + f$. Which of the following equals $a + b + c + d + e + f$?

(A) $2b$

(B) $2c$

(C) $2a + 2b$

(D) $2h$

(E) $2k$

47. (#13, AMC 12A, 2002) Two different positive numbers a and b each differ from their reciprocals by 1. What is $a + b$?

(A) 1

(B) 2

(C) $\sqrt{5}$

(D) $\sqrt{6}$

(E) 3

48. (#19, AMC 12A, 2003) A parabola with equation $y = ax^2 + bx + c$ is reflected about the x-axis. The parabola and its reflection are translated horizontally five units in opposite directions to become the graphs of $y = f(x)$ and $y = g(x)$, respectively. Which of the following describes the graph of $y = (f + g)(x)$?

(A) a parabola tangent to the x-axis

(B) a parabola not tangent to the x-axis

(C) a horizontal line

(D) a nonhorizontal line

(E) the graph of a cubic function

49. (#18, AMC 12B, 2004) Points A and B are on the parabola

$$y = 4x^2 + 7x - 1,$$

and the origin is the midpoint of $\overline{AB}$. What is the length of $\overline{AB}$?

(A) $2\sqrt{5}$

(B) $5 + \dfrac{\sqrt{2}}{2}$

(C) $5 + \sqrt{2}$

(D) 7

(E) $5\sqrt{2}$

50. (#21, AMC 12A, 2007) The sum of the zeros, the product of the zeros, and the sum of the coefficients of the function $f(x) = ax^2 + bx + c$ are equal. Their common value must be which of the following?
(A) the coefficient of x^2
(B) the coefficient of x
(C) the y-intercept of the graph of $y = f(x)$
(D) one of the x-intercepts of the graph of $y = f(x)$
(E) the mean of the x-intercepts of the graph of $y = f(x)$

51. (#14, AMC 12B, 2011) A segment though the focus F of a parabola with vertex V is perpendicular to $\overline{FV}$ and intersects the parabola in points A and B. What is $\cos(\angle AVB)$?
(A) $-\dfrac{3\sqrt{5}}{7}$
(B) $-\dfrac{2\sqrt{5}}{5}$
(C) $-\dfrac{4}{5}$
(D) $-\dfrac{3}{5}$
(E) $-\dfrac{1}{2}$

52. (#25, AMC 12A, 2003) Let $f(x) = \sqrt{ax^2 + bx}$. For how many real values of a is there at least one positive value of b for which the domain of f and the range of f are the same set?
(A) 0
(B) 1
(C) 2
(D) 3
(E) infinitely many

9

Polynomial Functions

Generalizing the notion of a quadratic expression, a *polynomial expression* in the variable x is any expression algebraically equivalent to one the form

$$a_n x^n + a_{n-1} x^{n-1} + \cdots + a_2 x^2 + a_1 x + a_0$$

for some nonnegative integer n and real (or complex) numbers a_n, a_{n-1}, ..., a_2, a_1, and a_0.

For example, each of the following equations defines a polynomial function in the variable x:

$$p(x) = 4x^5 - 6x^4 + 2x^3 + x^2 - x - 133,$$
$$q(x) = x^{100} - \frac{x}{100},$$
$$r(x) = x,$$
$$s(x) = 2x^2 + 3x + 1,$$
$$t(x) = 23.$$

The equation $h(x) = \frac{1}{x} + \frac{1}{x^2} + \frac{1}{x^3}$ does not define a polynomial function in the variable x (but it does in the variable $\frac{1}{x}$!).

Jargon. For a polynomial expression presented in the form $a_n x^n + a_{n-1} x^{n-1} + \cdots + a_1 x + a_0$ with $a_n \neq 0$, n is called the *degree* of the polynomial, a_n the *leading coefficient*, and a_0 the *constant term* of the polynomial.

For example, $2x - x^5$ is a degree-5 polynomial with leading term -1 and constant term 0.

A degree-0 polynomial is a constant, a degree-1 polynomial is a linear expression, and a degree-2 polynomial is a quadratic expression.

Comment. The work of "Exploding Dots" (see `www.gdaymath.com/courses/`) makes it clear that polynomial work is a generalization of base-10 place-value in arithmetic. For example, the number 2768 in base 10 is $2 \times 10^3 + 7 \times 10^2 + 6 \times 10 + 8 \times 1$. In base x it is $2x^3 + 7x^2 + 6x + 8$. The arithmetic of polynomials closely matches the arithmetic of numbers.

The Long-Term Behavior of Polynomial Functions

Let's first review some mathematical notation.

The symbol ∞ is used in mathematics as shorthand for the phrase "arbitrarily large and positive" and the symbol $-\infty$ means "arbitrarily large and negative". Also, an arrow $\to$ is often used as shorthand for the word "becomes" or "approaches". So a statement such as

$$x^2 \to 9 \text{ as } x \to 3$$

reads as follows: "The quantity x^2 approaches the value 9 as x approaches the value 3." This is a true statement. A table of values certainly suggests this is so.

$$
\begin{aligned}
x &= 2, & x^2 &= 4, \\
x &= 2.9, & x^2 &= 8.41, \\
x &= 2.99, & x^2 &= 8.9401, \\
x &= 2.9999, & x^2 &= 8.99940001, \\
x &= 3.5, & x^2 &= 12.25, \\
x &= 3.1, & x^2 &= 9.61, \\
x &= 3.01, & x^2 &= 9.0601, \\
x &= 3.0001, & x^2 &= 9.00060001.
\end{aligned}
$$

We can argue in general that if x adopts a value that is very close to 3, then x^2 will have a value very close to 9.

Mathematicians might also write this statement as

$$\lim_{x \to 3} x^2 = 9$$

to be read as follows: "The limit of x^2 as x approaches 3 is 9."

As another example

$$\lim_{x \to \infty} x^3 = \infty$$

reads as follows: "As x becomes large and positive, we have that x^3 becomes large and positive." This is indeed true.

Exercise. Interpret the statement

$$\lim_{x \to \infty} \frac{1}{x} = 0.$$

Solution. As x becomes large and positive, $\frac{1}{x}$ approaches the value 0. (This is true.) $\qquad\square$

Exercise. Interpret the statement

$$\lim_{x \to -\infty} x^4 = \infty.$$

Solution. As x becomes large and negative, x^4 becomes large and positive. (This is true.) $\qquad\square$

Exercise. Evaluate each of the following.

(a) $\lim_{x \to -\infty} \frac{1}{x^3}$

(b) $\lim_{x \to \infty} \frac{2-x}{5}$

(c) $\lim_{x \to \infty} 2 + \frac{1}{x} + \frac{1}{x^2} + \frac{1}{x^5}$

(d) $\lim_{x \to 1} 2 + \frac{1}{x} + \frac{1}{x^2} + \frac{1}{x^5}$

Solution. (a) 0. (As x becomes large and negative, $\frac{1}{x^3}$ approaches the value 0.)

(b) $-\infty$. (As x becomes large and positive, $\frac{2-x}{5}$ becomes large and negative. Try $x = 100000$.)

(c) 2. (As x becomes large and positive, $2 + \frac{1}{x} + \frac{1}{x^2} + \frac{1}{x^5}$ approaches the value $2 + 0 + 0 + 0 = 2$.)

(d) 5. (As x approaches the value 1, $2 + \frac{1}{x} + \frac{1}{x^2} + \frac{1}{x^5}$ approaches the value $2 + 1 + 1 + 1 = 5$.) $\qquad\square$

Now let's examine the long-term behavior of polynomial functions; that is, let's examine the nature of the outputs of a polynomial function for inputs that become arbitrarily large and positive and arbitrarily large and negative.

Example. Find $\lim_{x \to \infty} x^4 - 2x^3 - x^2 - x + 8$ and $\lim_{x \to -\infty} x^4 - 2x^3 - x^2 - x + 8$.

It might be hard to get a feel for matters by substituting in large values for x, say $x = 1000000$ or $x = -1000000$: the addition and subtraction of large numbers can be awkward. But we can bring clarity to the situation by bringing focus to the highest power of x that appears in the expression. Let's "pull out" this highest power and write

$$x^4 - 2x^3 - x^2 - x + 8 = x^4 \left(1 - \frac{2}{x} - \frac{1}{x^2} - \frac{1}{x^3} + \frac{8}{x^4} \right).$$

Now, as $x \to \infty$, the quantity

$$1 - \frac{2}{x} - \frac{1}{x^2} - \frac{1}{x^3} + \frac{8}{x^4}$$

approaches the value $1 + 0 - 0 + 0 + 0 = 1$ and the quantity x^4 becomes very large and positive. The net effect is that $x^4 \left(1 - \frac{2}{x} - \frac{1}{x^2} - \frac{1}{x^3} + \frac{8}{x^4} \right)$ becomes a large positive number multiplied by a number very close to 1. The result is a large positive number. We have

$$\lim_{x \to \infty} x^4 - 2x^3 - x^2 - x + 8 = \lim_{x \to \infty} x^4 \left(1 - \frac{2}{x} - \frac{1}{x^2} - \frac{1}{x^3} + \frac{8}{x^4} \right)$$
$$= \text{"big positive} \times 1\text{"}$$
$$= \infty.$$

In the same way, we see that

$$\lim_{x \to -\infty} x^4 - 2x^3 - x^2 - x + 8 = \lim_{x \to -\infty} x^4 \left(1 - \frac{2}{x} - \frac{1}{x^2} - \frac{1}{x^3} + \frac{8}{x^4} \right)$$
$$= \text{"big positive} \times 1\text{"}$$
$$= \infty.$$

Thus the value of the polynomial expression $x^4 + 2x^3 - x^2 + x + 8$ becomes large and positive as x becomes large and positive and also as x becomes large and negative.

Example. Explore the long-term behavior of $p(x) = 3x^3 - 2x^2 + 7x + 1$.

We have

$$\lim_{x \to \infty} 3x^3 - 2x^2 + 7x + 1 = \lim_{x \to \infty} x^3 \left(3 - \frac{2}{x} + \frac{7}{x^2} + \frac{1}{x^3} \right)$$
$$= \text{``big positive} \times 3\text{''}$$
$$= \infty,$$

$$\lim_{x \to -\infty} 3x^3 - 2x^2 + 7x + 1 = \lim_{x \to -\infty} x^3 \left(3 - \frac{2}{x} + \frac{7}{x^2} + \frac{1}{x^3} \right)$$
$$= \text{``big negative} \times 3\text{''}$$
$$= -\infty.$$

The outputs of the function $p(x) = 3x^3 - 2x^2 + 7x + 1$ become arbitrarily large and positive and arbitrarily large and negative. If we are convinced that the graph of this function has no holes or gaps (that the graph is sure to be a continuous curve), then we can be sure that the function $p(x) = 3x^3 - 2x^2 + 7x + 1$ will, for some inputs, give output values one million, one trillion, and one googleplex plus three. It will also give output values negative five, negative three billion, and negative ten raised to the three-hundred and thirty-third power.

Exercise. Discuss the long-term behaviors of the following polynomial functions:

(a) $a(x) = x^2 - x$.

(b) $b(x) = x^9 - 3x^5 + 4x^4 - 2x$.

(c) $c(x) = x^{25} - x^{50} - 1$.

Solution.   (a) $\lim_{x \to \infty} a(x) = \infty$; $\lim_{x \to -\infty} a(x) = \infty$.

(b) $\lim_{x \to \infty} b(x) = \infty$; $\lim_{x \to -\infty} b(x) = -\infty$.

(c) $\lim_{x \to \infty} c(x) = -\infty$; $\lim_{x \to -\infty} c(x) = -\infty$.   □

Exercise. Can we be sure that the polynomial p given by $p(x) = 5x^6 + 8x^4 - 7x^3 - 19x^2 + 38x - 100$ gives outputs of any specified negative value, say, negative ten thousand or negative thirteen million?

Solution. Since $\lim_{x \to \infty} p(x) = \infty$ and $\lim_{x \to -\infty} p(x) = \infty$, we can't be sure that p adopts arbitrarily large and negative output values.   □

Exercise. Is it possible for a polynomial p to have the property that $\lim_{x \to \infty} p(x) = 17$?

Must a polynomial always approach ∞ or $-\infty$ in its long-term behavior?

Solution. If $p(x) = a_n x^n + a_{n-1} x^{n-1} + \cdots + a_1 x + a_0$ with $n \geq 1$, then writing $p(x)$ as

$$p(x) = x^n \left(a_n + \frac{a_{n-1}}{x} + \cdots + \frac{a_1}{x^{n-1}} + \frac{a_0}{x^n} \right)$$

shows that we must have $\lim_{x \to \infty} p(x) = \pm\infty$ and $\lim_{x \to -\infty} p(x) = \pm\infty$. The only way we can have $\lim_{x \to \infty} p(x) = 17$ is for p to be the constant polynomial $p(x) = 17$. $\square$

The Graphs of $y = x^n$

Let's consider the very basic polynomials equations of the form $y = x^n$ and their graphs.

Cases $n = 0$ and $n = 1$.

The graph of $y = 1$ is a horizontal line, and of $y = x$ a diagonal line.

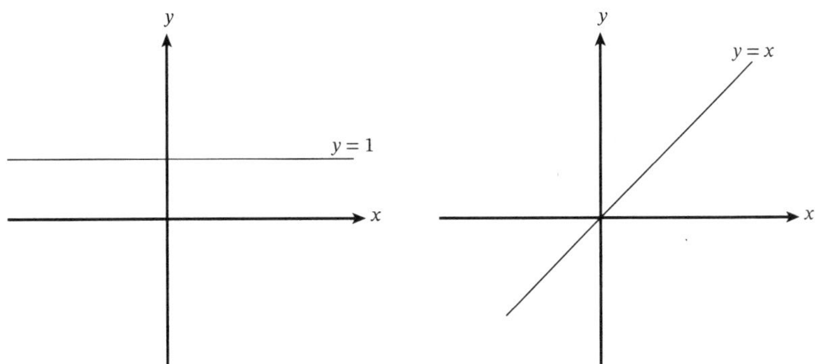

FIGURE 57

Aside. Here's an annoying question: Is the graph of $y = x^0$ defined at $x = 0$? Should the left graph of Figure 57 show a gap at the point $(0, 1)$?

Cases $n = 2$ and n even.

The case $n = 2$ is the quadratic function $y = x^2$. See Figure 58.

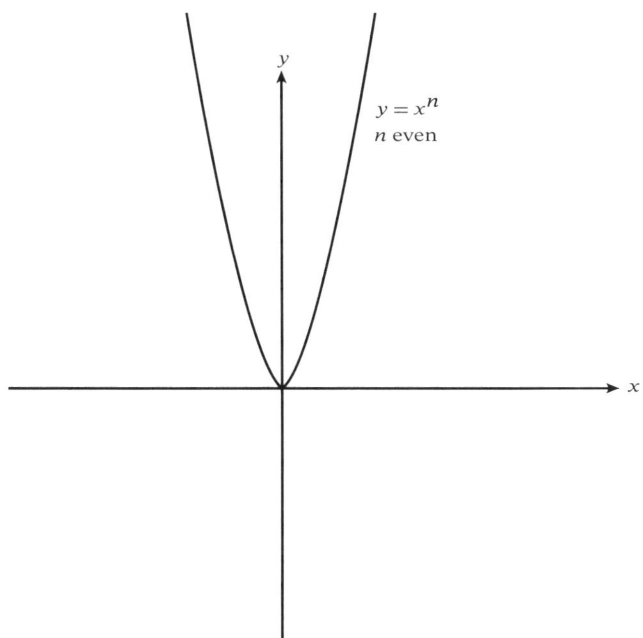

FIGURE 58

For any even value of n, the equation $y = x^n$ has the same long-term behavior as $y = x^2$:

$$\lim_{x \to \infty} x^n = \infty,$$

$$\lim_{x \to -\infty} x^n = \infty.$$

By plotting points, one sees that its graph has the same basic U-shape as the squaring function, except that it rises to higher values more quickly. (The larger the value n the tighter the U-shape.)

Cases $n = 3$ and n odd.

The case $n = 3$ produces the cubing function $y = x^3$ with long-term behavior

$$\lim_{x \to \infty} x^3 = \infty,$$

$$\lim_{x \to -\infty} x^3 = -\infty.$$

An exercise in plotting points shows that its graph has an S-bend shape.
See Figure 59.

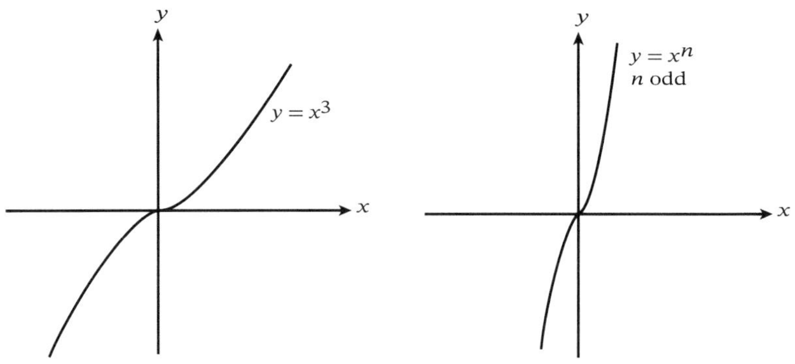

FIGURE 59

In general, for odd n, the function given by $y = x^n$ has the same
long-term behavior:

$$\lim_{x \to \infty} x^n = \infty,$$

$$\lim_{x \to -\infty} x^n = -\infty,$$

and, by plotting points, one sees that its graph has the same basic S-shape
as the cubing function, except tighter for higher values of n.

Aside. Consider the equation $y = x^{\frac{1}{n}}$. Algebra shows that this can be
rewritten $x = y^n$ and so the graph of $y = x^{\frac{1}{n}}$ is the same as the graph of
$y = x^n$, but with the x- and y-axes reversed. Bringing the x-axis back to
horizontal and the y-axis to vertical, the net effect is a reflection about
the diagonal line $y = x$. See Figure 60.

Here we are assuming that $y = x^{\frac{1}{n}}$ is defining a single-valued function, setting the output to be only the positive root of the input x if there
is a choice between a positive or negative root.

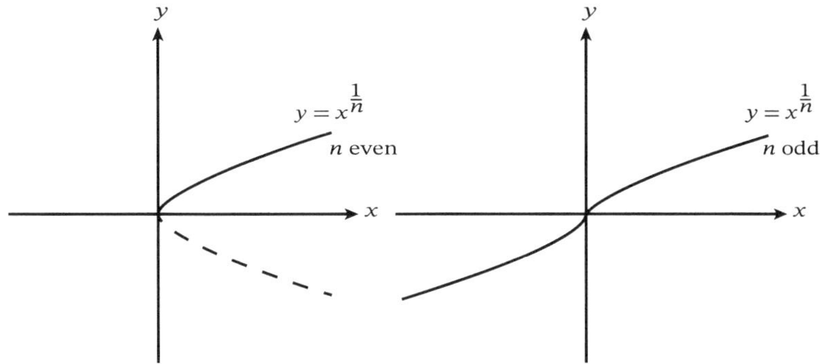

FIGURE 60

Graphs of Factored Polynomials

Consider a polynomial expression that is factored into a product of linear terms. (Some linear terms may be repeated.) As a specific example, let's consider the polynomial p given by

$$p(x) = 5(x + 6)^3(x + 1)^2(x - 2)(x - 3)^8(x - 4)^6$$

and see if we can graph the function.

If we were to expand the brackets, we'd see that p is a polynomial of degree 20 with leading coefficient 5. Its long-term behavior would thus be

$$\lim_{x \to \infty} p(x) = \infty,$$

$$\lim_{x \to -\infty} p(x) = \infty.$$

One can also see this by imagining placing a large positive and a large negative number for x into $p(x)$:

$$p(\text{positve big}) = 5 \times \text{positive big} \times \text{positive big} \times \text{positive big}$$
$$\times \text{positive big} \times \text{positive big}$$
$$= \text{positive big},$$

$$p(\text{negative big}) = 5 \times \text{negative big} \times \text{positive big} \times \text{negative big}$$
$$\times \text{positive big} \times \text{positive big}$$
$$= \text{positive big}.$$

Let's draw this. See Figure 61.

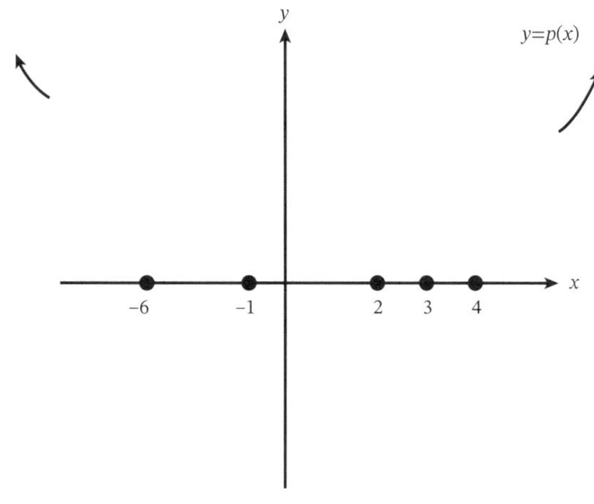

FIGURE 61

Also, that the polynomial is factored makes it clear that $p(x) = 0$ for $x = -6, -1, 2, 3,$ and 4. We've added that information to Figure 61. No other value for x will give $p(x) = 0$.

Ah! The graph of the polynomial touches the x-axis at no other values. So between $x = -6$ and $x = -1$, say, the graph of the function is either entirely above the x-axis or below. We can test which by evaluating the polynomial at some input between the two. We have for $x = -3$, for instance,

$$p(-3) = 5 \times (\text{positive})^3 \times (\text{negative})^2 \times (\text{negative}) \times (\text{negative})^8 \times (\text{negative})^4,$$

which is a negative value. The graph of the function is below the axis for x between -6 and -1.

Examining $p(0)$, for instance, shows that $p(x)$ is also negative for x between -1 and 2:

$$p(0) = 5 \times (\text{positive})^3 \times (\text{positive})^2 \times (\text{negative}) \times (\text{negative})^8 \times (\text{negative})^4$$
$$= \text{negative}.$$

We also have

$$p(2.5) = 5 \times (\text{positive})^3 \times (\text{positive})^2 \times (\text{positive}) \times (\text{negative})^8 \times (\text{negative})^4$$
$$= \text{positive},$$

showing that $p(x)$ is positive for x between 2 and 3. And an examination of $p(3.7)$, say, shows that $p(x)$ is also positive for x between 3 and 4.

The long-term behavior of the polynomial also establishes that $p(x)$ is positive for all $x < -6$ and also for all $x > 4$. Figure 62 records this information.

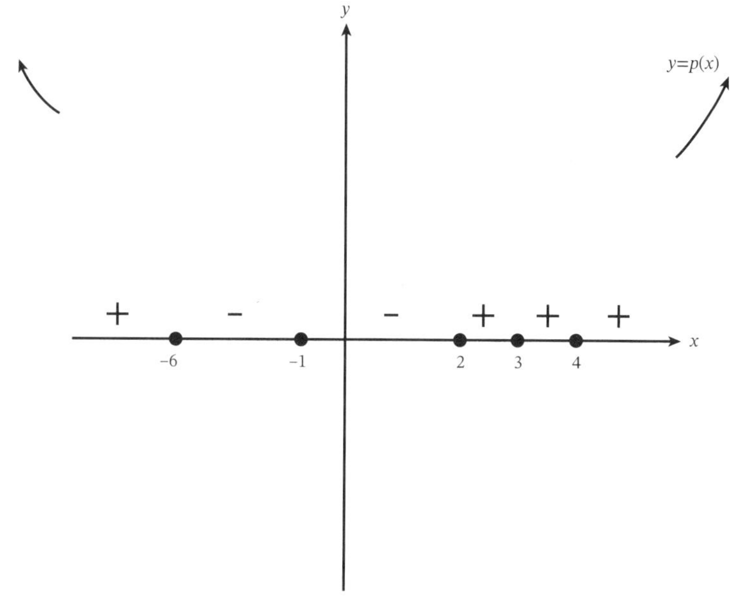

FIGURE 62

This now gives a sense of how the entire graph of the polynomial might appear.

Figure 63 is a crude sketch. For example, the graph of the polynomial might, for all we know, actually appear as follows as in Figure 64, with multiple "dips and humps" between the zeros.

With the tools of calculus one can garner a better understanding of the true shapes of polynomial graphs. It turns out that the graph of a degree-n polynomial will have at most $n - 1$ dips and humps. But given the theory we have developed thus far, we have no means to know this.

As one practices analyzing the behavior of polynomial graphs, one might come to notice the following.

If a linear term $x - a$ in a factored expression of a polynomial appears raised to an odd degree, $(x - a)^{\text{odd}}$, then just to the left of $x = a$ and just to the right of $x = a$, polynomial outputs will have different signs and

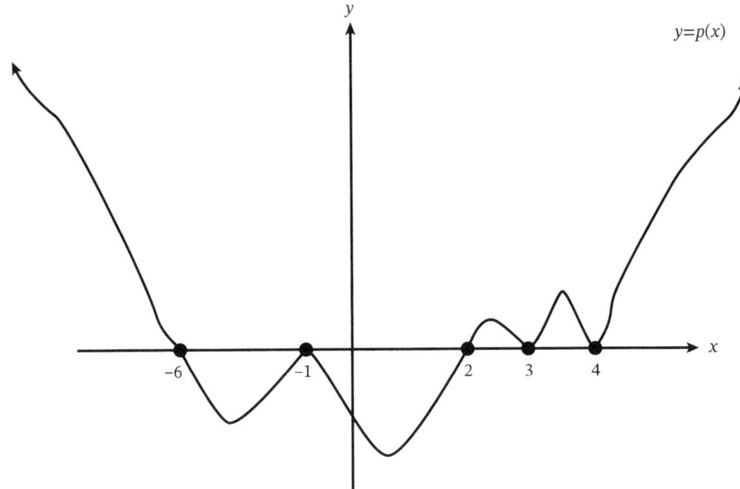

FIGURE 63

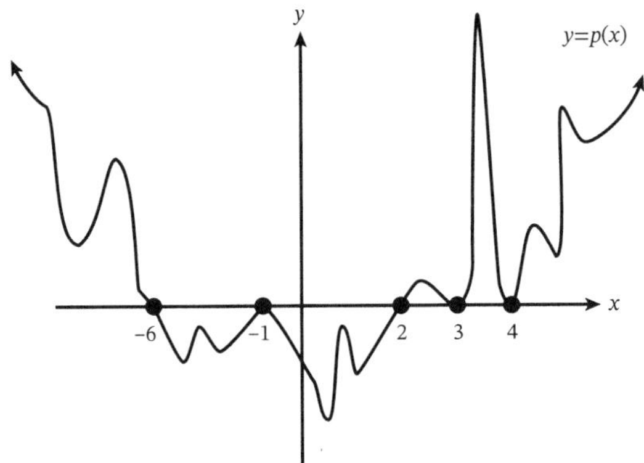

FIGURE 64

so its graph crosses the axis at $x = a$. (There is a change of sign either side of the factor $(x - a)^{\text{odd}}$, but no other factor gives a change of sign.)

For example, for $p(x) = 5(x + 6)^3(x + 1)^2(x - 2)(x - 3)^8(x - 4)^6$ compare the signs of $p(-6.1)$ and $p(-5.9)$.

If a linear term $x - a$ in a factored expression of a polynomial appears raised to an even degree, $(x - a)^{\text{even}}$, then the sign of outputs on either side of $x = a$ do not change and so the polynomial just touches the axis at $x = a$. (Neither $(x - a)^{\text{even}}$ nor any other factor yields a change of sign.)

For example, for $p(x) = 5(x + 6)^3(x + 1)^2(x - 2)(x - 3)^8(x - 4)^6$ compare the signs of $p(2.9)$ and $p(3.1)$.

This observation, along with analysis of the long-term behavior of a polynomial, makes sketching graphs of factored polynomials remarkably speedy.

Example. Make a crude sketch of $q(x) = (x + 4)^2(x + 1)^3 x^4 (x - 3)^8$.

This is a degree-17 polynomial with leading coefficient 1. Examination of q(big and positive) and q(big and negative) shows us that

$$\lim_{x \to \infty} q(x) = \infty,$$

$$\lim_{x \to -\infty} q(x) = -\infty.$$

The graph has zeros $x = -4$ (just touching), $x = -1$ (crossing), $x = 0$ (just touching), $x = 3$ (just touching). We must have a graph of the form of Figure 65.

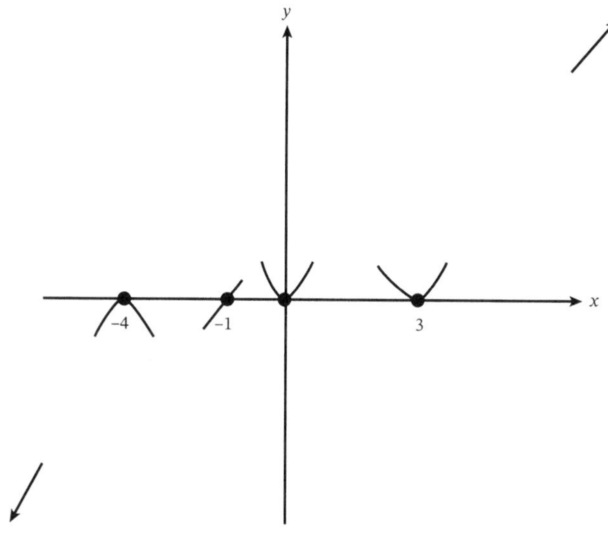

FIGURE 65

We can now fill in the gaps to complete a sketch of the graph.

Comment. Some textbook authors insist that students add an additional feature of detail: if a factored polynomial expression has a term of the form $(x - a)^n$, with n an odd integer 3 or greater, then the graph of the polynomial crosses the x-axis at $x = a$ just as the graph of $y = x^n$ crosses the axis, namely as an S-shape.

Figure 66 shows this additional detail.

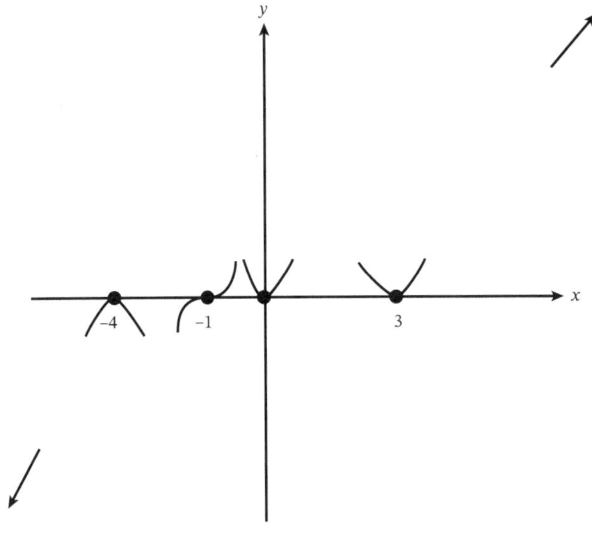

FIGURE 66

It is a matter of taste as to how pedantic one wishes to be. Given that one can't garner (or prove) any true detail about these polynomial graphs without the aid of calculus, one might opt for less fussiness.

Graphs of Nonfactored Polynomials

Not all polynomials factor into linear terms. For example,

$$p(x) = (x^2 + 1)(x^2 + 2)$$

does not factor further. Its long-term behavior is clear:

$$\lim_{x \to \infty} p(x) = \infty,$$

$$\lim_{x \to -\infty} p(x) = \infty,$$

but this function has no (real) zeros and so its graph does not cross the x-axis. Our methods for sketching its graph fail us.

Without calculus, we must return to basic principles and simply plot points (just as the graphing calculator does!).

Sometimes students might be given a polynomial that can be factored, but in nonfactored form. Students are then expected to use the techniques of factoring taught in algebra class to rewrite the polynomial expression in factored form in order to be able to graph the function. Factoring a polynomial is usually no easy task!

But there are some theoretical results to help identify factors.

The Factor Theorem

One can identify a factor of a polynomial by locating a zero of the polynomial function. The famous factor theorem states:

> If a polynomial p has a zero at $x = a$ (so $p(a) = 0$), then $p(x)$ can be rewritten as
>
> $$p(x) = (x - a) \times (\text{another polynomial}).$$
>
> The second polynomial that arises has degree 1 less than the degree of p.

One establishes this theorem by first looking at polynomial division. Consider, for example

$$\frac{2x^2 + 7x + 3}{x - 1},$$

the polynomial $p(x) = 2x^2 + 7x + 3$ divided by $x - 1$.

Is $2x^2 + 7x + 3$ a multiple of $x - 1$? Is some portion of it such a multiple?

Start with the $2x^2$ term. This is a multiple of x, namely $2x(x)$, which is close to being a multiple of $x - 1$. Let's arrange matters so that $2x(x - 1) = 2x^2 - 2x$ appears:

$$\frac{2x^2 + 7x + 3}{x - 1} = \frac{2x^2 - 2x + 2x + 7x + 3}{x - 1}$$
$$= \frac{2x(x - 1) + 9x + 3}{x - 1}.$$

The next term to consider is $9x$. This is a multiple of x. Can we adjust matters so that $9(x - 1)$, a multiple of $x - 1$, appears instead?

$$\frac{2x(x-1) + 9x + 3}{x-1} = \frac{2x(x-1) + 9x - 9 + 9 + 3}{x-1}$$
$$= \frac{2x(x-1) + 9(x-1) + 12}{x-1}.$$

Now it is clear that

$$\frac{2x^2 + 7x + 3}{x-1} = 2x + 9 + \frac{12}{x-1}.$$

Can we make sense of the number 12 that appeared as the remainder?

We wrote $p(x) = 2x^2 + 7x + 3$ as $p(x) = 2x(x-1) + 9(x-1) + 12$. It seems irresistible to put in $x = 1$. Doing so shows $p(1) = 0 + 0 + 12 = 12$. The remainder upon division by $x - 1$ is actually $p(1)$.

Exercise. Write $p(x) = x^3 - 2x + 1$ as a multiple of $x - 5$, if possible. If it is not possible, get close!

Solution. We have

$$\begin{aligned}
x^3 - 2x + 1 &= x^2(x - 5) + 5x^2 - 2x + 1 \\
&= x^2(x - 5) + 5x(x - 5) + 23x + 1 \\
&= x^2(x - 5) + 5x(x - 5) + 23(x - 5) + 116.
\end{aligned}$$

We see that $p(5) = 116$. $\square$

This approach shows that given a polynomial p and a linear term $x - a$, we can always write $p(x)$ as multiples of $x - a$ plus a single number r:

$$p(x) = (\text{multiples of } x - a) + r.$$

Putting in $x = a$ shows that $p(a) = r$. Consequently,

$$p(x) \text{ is evenly divisible by } x - a \text{ if and only if } p(a) = 0.$$

That is,

$$\text{If } x = a \text{ is a zero of } p, \text{ then } x - a \text{ is a factor of } p(x).$$

A comment on complex roots

Suppose a polynomial p has real coefficients, $p(x) = a_n x^n + \cdots + a_1 x + a_0$ with each a_i real, and suppose $x = a + ib$ is a complex number root of the polynomial. This means that

$$a_n(a + ib)^n + \cdots + a_1(a + ib) + a_0 = 0.$$

Taking the complex conjugate of the entire equation (and using the fact that $\overline{z^k} = (\overline{z})^k$) we see that

$$a_n(a - ib)^n + \cdots + a_1(a - ib) + a_0 = 0$$

as well. That is, $\overline{x}$ is also a root of the equation. This shows that *all complex roots of polynomials with real coefficients come in conjugate pairs.*

By the factor theorem (whose proof is valid even in complex number arithmetic), we see then that both $(x - a - ib)$ and $(x - a + ib)$ are factors of $p(x)$. Consequently, $p(x)$ has the (real) quadratic factor

$$(x - a - ib)(x - a + ib) = (x - a)^2 + b^2.$$

One can prove in advanced mathematics that a polynomial of degree n has precisely n roots, though some may be complex roots, and some values may be repeated. If a polynomial has odd degree, then it must have at least one real root as all complex roots come in conjugate pairs.

A comment on rational roots

If a polynomial P given by $P(x) = a_n x^n + a_{n-1} x^{n-1} + \cdots + a_1 x + a_0$ has integer coefficients, $a_n \neq 0$, and $x = \frac{p}{q}$ is a rational root of the polynomial, with p and q integers sharing no common factor, then multiplying the equation $a_n \left(\frac{p}{q} \right)^n + a_{n-1} \left(\frac{p}{q} \right)^{n-1} + \cdots + a_1 \left(\frac{p}{q} \right) + a_0 = 0$ through by q^n gives the equation

$$a_n p^n + a_{n-1} p^{n-1} q + \cdots + a_1 p q^{n-1} + a_0 q^n = 0.$$

This shows that $a_n p^n$ equals a multiple of q. Since p^n shares no common factors with q, it follows that a_n must be a multiple of q. Similarly we see that $a_0 q^n$ equals a multiple of p, from which we conclude that a_0 is a multiple of p. We have the *rational roots theorem*:

> Any rational root $x = \frac{p}{q}$, written in reduced form, of a polynomial with integer coefficients must have p a factor of the constant coefficient and q a factor of the leading coefficient.

Consequently, any polynomial with integer coefficients and leading coefficient 1 only has integer and irrational real zeros (if any).

The rational roots theorem allows us to identify possible roots of a polynomial and hence possible factors of the polynomial.

Exercise. Are there any rational solutions to $2x^4 - 3x^3 + 7x + 3 = 0$?

Solution. If $x = \frac{p}{q}$ is a rational solution written in reduced form, then p is a factor of 3 and so is either 1, -1, 3, or -3, and q is a factor of 2 and so is either 1, -1, 2, or -2. This gives the following options for x:

$$1, -1, \frac{1}{2}, -\frac{1}{2}, \frac{3}{2}, -\frac{3}{2}, 3, -3.$$

Checking each in turn, we see that $x = -\frac{1}{2}$ is a solution to the equation. (And this is the only rational solution to the equation.) $\qquad\square$

MAA Featured Problem

(#23, AMC 12A, 2004)

A polynomial

$$P(x) = c_{2004}x^{2004} + c_{2003}x^{2003} + \cdots + c_1 x + c_0$$

has real coefficients with $c_{2004} \neq 0$ and 2004 distinct complex zeros $z_k = a_k + b_k i$, $1 \leq k \leq 2004$, with a_k and b_k real, $a_1 = b_1 = 0$, and

$$\sum_{k=1}^{2004} a_k = \sum_{k=1}^{2004} b_k.$$

Which of the following quantities can be a nonzero number?

    (A) c_0
    (B) c_{2003}
    (C) $b_2 b_3 \cdots b_{2004}$
    (D) $\sum_{k=1}^{2004} a_k$
    (E) $\sum_{k=1}^{2004} c_k$

A Personal Account of Solving This Problem

Curriculum Inspirations Strategy (`www.maa.org/ci`):

> Strategy 6: Eliminate incorrect choices.

This question looks positively nightmarish!

I feel like I first need to collect my thoughts about what I know about roots of polynomials.

Some facts:

> If $x = r$ is a zero of a polynomial, then $x - r$ is a factor of the polynomial.

> If $x = a + bi$ is a zero of a polynomial with real coefficients, then so is $\overline{x} = a - bi$. (That is, nonreal roots come in conjugate pairs.)

> A polynomial of degree 2004 has at most that many distinct roots.

Anything else? There is probably more, but let's get started on the question.

The polynomial in the question has real coefficients and has distinct roots $a_1 + b_1 i = 0, a_2 + b_2 i, a_3 + b_3 i, \ldots$.

Since $P(0) = 0$, we have that $c_0 = 0$. Option (A) is out.

Option (E) is the sum of the coefficients of the polynomial. That's $P(1)$. Could $P(1)$ be 0? I don't know.

Hmm. The complex roots of the polynomial come in conjugate pairs. We have one real root, $a_1 + b_1 i = 0$, and the degree of the polynomial is 2004. There must be another real root. That means some other b_i is also 0 and so the product $b_2 b_3 \cdot \cdots \cdot b_{2004}$ is 0. Option (C) is out.

Oh! The remaining roots come in conjugate pairs and so $\sum_{k=1}^{2004} b_k = 0$. Since this sum, we are told, equals $\sum_{k=1}^{2004} a_k = 0$, option (D) is out.

It's now between (B) and (E).

Must c_{2003} be 0? Or must $\sum_{k=1}^{2004} c_k = P(1)$ be 0?

We haven't used yet the fact that if $x = r$ is a zero of a polynomial, then $x - r$ is a factor of the polynomial. We have

$$P(x) = c_{2004} x (x - a_1 - b_1 i) \cdot \cdots \cdot (x - a_{2004} - b_{2004} i).$$

I don't see this as helpful. If we expanded this, we'd see that c_0 is 0. But we already knew that.

Oh! We'd also see that

$$p(x) = c_{2004} x^{2004}$$
$$+ x^{2003}(-a_1 - b_i i - a_2 - b_2 i - \cdots - a_{2004} - b_{2004} i) + \cdots.$$

So

$$c_{2003} = -\sum_{k=1}^{2004} a_k - i \sum_{k=1}^{2004} b_k = -0 - 0i = 0.$$

This rules out (B) and so the answer is (E).

(We should give an explicit example of a polynomial that satisfies all the properties given in the question and definitely, for sure, has $P(1) \neq 0$, to show that this competition question is meaningful. Can you find one?)

Additional Problems

53. (#3, AMC 10B, 2002) For how many positive integers n is $n^2 - 3n + 2$ a prime number?

(A) none

(B) one

(C) two

(D) more than two, but finitely many

(E) infinitely many

54. (#17, AMC 12B, 2004) For some real numbers a and b, the equation

$$8x^3 + 4ax^2 + 2bx + a = 0$$

has three distinct positive roots. If the sum of the base-2 logarithms of the roots is 5, what is the value of a?

(A) -256

(B) -64

(C) -8

(D) 64

(E) 256

55. (#18, AMC 12A, 2007) The polynomial

$$f(x) = x^4 + ax^3 + bx^2 + cx + d$$

has real coefficients, and $f(2i) = f(2+i) = 0$. What is $a + b + c + d$?

(A) 0

(B) 1

(C) 4

(D) 9

(E) 16

56. (#21, AMC 12A, 2003) The graph of the polynomial

$$P(x) = x^5 + ax^4 + bx^3 + cx^2 + dx + e$$

has five distinct x-intercepts, one of which is at $(0, 0)$. Which of the following coefficients cannot be 0?

(A) a

(B) b

(C) c

(D) d

(E) e

57. (#23, AMC 12, 2001) A polynomial of degree 4 with leading coefficient 1 and integer coefficients has two real zeros, both of which are integers. Which of the following can also be a zero of the polynomial?

(A) $\dfrac{1 + i\sqrt{11}}{2}$

(B) $\dfrac{1 + i}{2}$

(C) $\dfrac{1}{2} + i$

(D) $1 + \dfrac{i}{2}$

(E) $\dfrac{1 + i\sqrt{13}}{2}$

58. (#23, AMC 12B, 2004) The polynomial $x^3 - 2004x^2 + mx + n$ has integer coefficients and three distinct positive zeros. Exactly one of these is an integer, and it is the sum of the other two. How many values of n are possible?

(A) 250,000

(B) 250,250

(C) 250,500

(D) 250,750

(E) 251,000

59. (#25, AMC 12A, 2002) The nonzero coefficients of a polynomial P with real coefficients are all replaced by their mean to form a polynomial Q. Which of the following could be a graph of $y = P(x)$ and $y = Q(x)$ over the interval $-4 \le x \le 4$?

(A) 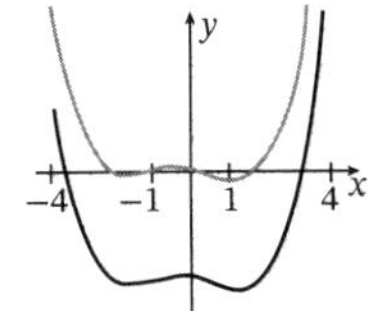

(B) 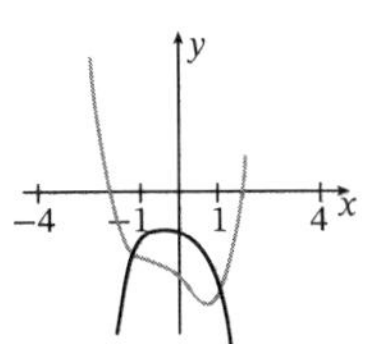

(C) 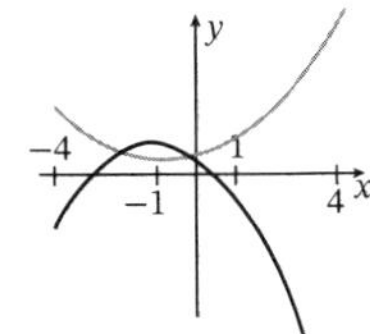

(D) 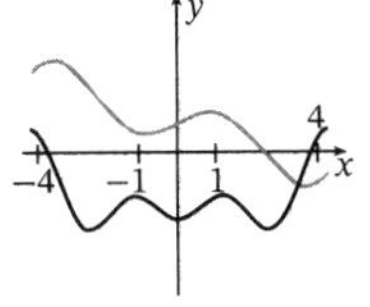

(E) 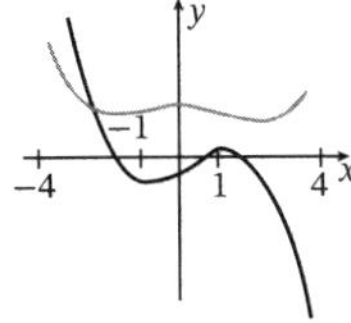

10

Rational Functions

Although the graphing of rational functions does not generally appear as a topic in the AMC, we include a discussion of it here for the sake of completeness.

Generalizing the notion of a polynomial expression, a *rational expression* in a variable x is any expression algebraically equivalent to one of the form $\frac{p(x)}{q(x)}$ with p and q polynomial expressions in x (and q not the polynomial that is identically 0).

For example, $\frac{x^2-2x+1}{(x-1)(x-4)^2}$, $\frac{1}{2x-7}$, and x^2, which is equivalent to $\frac{x^2}{1}$, and 1, which is equivalent to $\frac{1}{1}$, are each rational expressions in the variable x.

A *rational function* is a function whose outputs for an input x are given by a formula that is a rational expression of x. As such, a rational function is not defined at any values x that give a denominator of 0 in that rational expression.

For example, the rational function f defined by

$$f(x) = \frac{(x-3)^2(x^2+4)(x+7)^2}{5(x-2)(x-4)}$$

is defined only for values x different from 2 and 4.

There are subtle issues afoot when simplifying rational expressions by canceling common terms from the numerator and the denominator of the expression. Consider, for instance, the function g given by

$$g(x) = \frac{x^2-4}{x-2}.$$

This function is not defined at $x = 2$. Algebraically, however, we may write

$$g(x) = \frac{x^2 - 4}{x - 2}$$
$$= \frac{(x + 2)(x - 2)}{x - 2}$$
$$= x + 2$$

with the cancellation being valid only for those values of x with $x - 2$ nonzero, that is, with $x \neq 2$. Thus we have

$$g(x) = \frac{x^2 - 4}{x - 2} = \begin{cases} x + 2 & \text{if } x \neq 2, \\ \text{undefined} & \text{if } x = 2. \end{cases}$$

Its graph matches the graph of the straight line $y = x + 2$ except at $x = 2$ where it is undefined. See Figure 67.

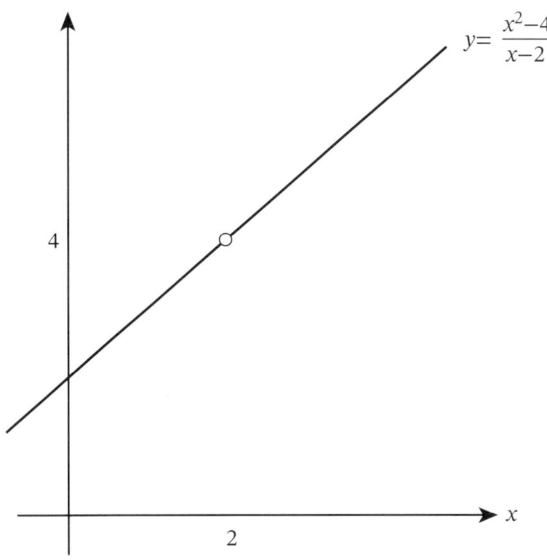

$$\text{F\scriptsize IGURE } 67$$

The graph of $y = \dfrac{(x-3)(x-5)(x-1)}{(x-1)}$ is a parabola with a hole at $x = 1$.

Exercise. Describe the graph of $y = \dfrac{x}{x}$.

Solution. It is a horizontal line one unit above the x-axis with a hole at $(0, 1)$. $\qquad\qquad\square$

Long-Term Behavior of Rational Functions

The long-term behavior of a polynomial function was best analyzed by pulling out the highest power of x that appears in the polynomial. The same device works well if we apply it separately to the numerator and to the denominator of the defining expression for a rational function.

Example. Analyze the long-term behavior of f given by

$$f(x) = \frac{4x^3 - 5x^2 + 7x + 9}{2x^3 - 5x + 13}.$$

Rewrite the numerator and denominator of the expression to highlight the highest power of x in each:

$$\frac{4x^3 - 5x^2 + 7x + 9}{2x^3 - 5x + 13} = \frac{x^3\left(4 - \dfrac{5}{x} + \dfrac{7}{x^2} + \dfrac{9}{x^3}\right)}{x^3\left(2 - \dfrac{5}{x^2} + \dfrac{13}{x^3}\right)}.$$

Since we are considering the long-term behavior of this function, that is, as x becomes arbitrarily large and positive or arbitrarily large and negative, we can assume x is not 0. Thus it is legitimate to cancel the common factor of x^3 that appears. We have

$$\lim_{x\to\infty} \frac{4x^3 - 5x^2 + 7x + 9}{2x^3 - 5x + 13} = \lim_{x\to\infty} \frac{x^3\left(4 - \dfrac{5}{x} + \dfrac{7}{x^2} + \dfrac{9}{x^3}\right)}{x^3\left(2 - \dfrac{5}{x^2} + \dfrac{13}{x^3}\right)}$$

$$= \lim_{x\to\infty} \frac{4 - \dfrac{5}{x} + \dfrac{7}{x^2} + \dfrac{9}{x^3}}{2 - \dfrac{5}{x^2} + \dfrac{13}{x^3}}.$$

We can now see that this limit has value $\dfrac{4-0+0+0}{2-0+0} = 2$. The same is true for the limit as $x \to -\infty$.

We conclude from this that the graph of the function

$$y = \frac{4x^3 - 5x^2 + 7x + 9}{2x^3 - 5x + 13}$$

for larger and larger positive and negative values of x approaches the value 2.

At this stage we have no idea in what manner its graph approaches this height of 2. It could be any number of ways, as suggested by Figure 68.

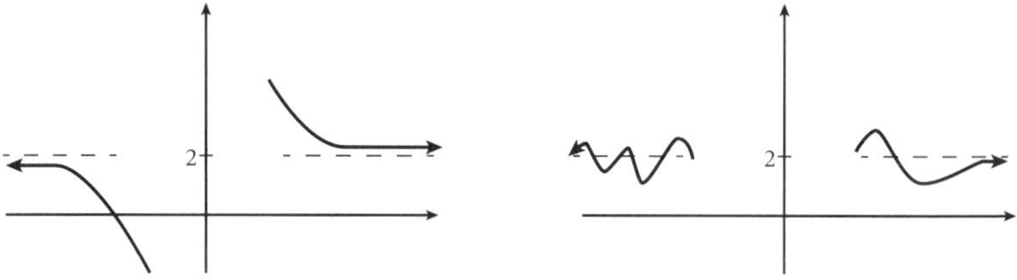

FIGURE 68

Jargon. Loosely speaking, we say a straight line is a (linear) *asymptote* for a graph $y = f(x)$ if the graph becomes arbitrarily close to that line for arbitrarily large values of x. The graph may cross that asymptotic line, many times in fact, but its distance away from that line must go to 0 in the long run. (Many people are confused by this: an asymptote may be crossed!)

For our example we see that the function $f(x) = \frac{4x^3 - 5x^2 + 7x + 9}{2x^3 - 5x + 13}$ has a horizontal asymptote to the right and to the left, and in both cases it is the horizontal line $y = 2$.

Exercise. Analyze the long-term behavior of $g(x) = \frac{x^4 - 1}{x^4 + x^2 + x}$.

Solution. We have

$$
\lim_{x \to \infty} \frac{x^4 - 1}{x^4 + x^2 + x} = \lim_{x \to \infty} \frac{x^4 \left(1 - \frac{1}{x^4}\right)}{x^4 \left(1 + \frac{1}{x^2} + \frac{1}{x^3}\right)}
$$

$$
= \lim_{x \to \infty} \frac{1 - \frac{1}{x^4}}{1 + \frac{1}{x^2} + \frac{1}{x^3}}
$$

$$
= \frac{1 - 0}{1 + 0 + 0}
$$

$$
= 1
$$

and

$$\lim_{x \to -\infty} \frac{x^4 - 1}{x^4 + x^2 + x} = \lim_{x \to -\infty} \frac{x^4\left(1 - \frac{1}{x^4}\right)}{x^4\left(1 + \frac{1}{x^2} + \frac{1}{x^3}\right)}$$

$$= \lim_{x \to -\infty} \frac{1 - \frac{1}{x^4}}{1 + \frac{1}{x^2} + \frac{1}{x^3}}$$

$$= \frac{1 - 0}{1 + 0 + 0}$$

$$= 1.$$

The function g has as horizontal asymptote the line $y = 1$ (both to the left and to the right). $\qquad\qquad\square$

Exercise. Analyze the long-term behavior of

$$h(x) = \frac{37x^5 - 87x^4 + 52x^3 - x^2 - 776x - 9987}{79x^5 + 652x^4 - 8x^3 - \frac{1}{2}x^2 + 6552x - 9}.$$

Solution. This function has as horizontal asymptote the line $y = \frac{37}{79}$ (both to the left and to the right). $\qquad\qquad\square$

Working through the previous exercise makes this next observation clear.

Suppose a rational function f has numerator and denominator polynomials of the same degree n, with leading coefficients a (for the numerator) and b (for the denominator):

$$f(x) = \frac{ax^n + \cdots}{bx^n + \cdots}.$$

Then f approaches the horizontal line $y = \frac{a}{b}$ for larger and larger positive and negative values of x.

Example. Analyze the long-term behavior of $f(x) = \frac{x^2 + 2x + 3}{x^3 + x^2 + 4x + 1}$.

Notice here that the numerator and denominator of this rational expression are polynomials of different degrees. Even so, let's apply the same algebraic technique to the expression: highlight the highest power of x that appears in each polynomial,

$$\lim_{x\to\infty} \frac{x^2 + 2x + 3}{x^3 + x^2 + 4x + 1} = \lim_{x\to\infty} \frac{x^2\left(1 + \frac{2}{x} + \frac{3}{x^2}\right)}{x^3\left(1 + \frac{1}{x} + \frac{4}{x^2} + \frac{1}{x^3}\right)}$$

$$= \lim_{x\to\infty} \frac{1}{x} \cdot \frac{1 + \frac{2}{x} + \frac{3}{x^2}}{1 + \frac{1}{x} + \frac{4}{x^2} + \frac{1}{x^3}}.$$

Now, as x grows, it is clear that $\frac{1}{x} \to 0$ and that

$$\frac{1 + \frac{2}{x} + \frac{3}{x^2}}{1 + \frac{1}{x} + \frac{4}{x^2} + \frac{1}{x^3}} \to \frac{1 + 0 + 0}{1 + 0 + 0 + 0} = 1.$$

We see that

$$\lim_{x\to\infty} \frac{x^2 + 2x + 3}{x^3 + x^2 + 4x + 1} = 0 \cdot 1 = 0.$$

In the same way, we have that

$$\lim_{x\to-\infty} \frac{x^2 + 2x + 3}{x^3 + x^2 + 4x + 1} = 0 \cdot 1 = 0.$$

This function has the x-axis as a horizontal asymptote both to the left and to the right.

We have the following:

Suppose a rational function f has the form

$$f(x) = \frac{ax^n + \cdots}{bx^m + \cdots}$$

with $m > n$. Then the horizontal line $y = 0$ is an asymptote for the function both to the left and to the right.

Example. Analyze the long-term behavior of

$$f(x) = \frac{x^5 - x^2 + 1}{2x^2 + 3}.$$

Let's apply the same technique:

$$\lim_{x \to \infty} \frac{x^5 - x^2 + 1}{2x^2 + 3} = \lim_{x \to \infty} \frac{x^5 \left(1 - \frac{1}{x^3} + \frac{1}{x^5}\right)}{x^2 \left(2 + \frac{3}{x^2}\right)}$$

$$= \lim_{x \to \infty} x^3 \cdot \frac{1 - \frac{1}{x^3} + \frac{1}{x^5}}{2 + \frac{3}{x^2}}.$$

Now as x grows, x^3 becomes large and positive, and the quantity

$$\frac{1 - \frac{1}{x^3} + \frac{1}{x^5}}{2 + \frac{3}{x^2}}$$

approaches the value $\frac{1}{2}$. Thus the function wants to become

$$(\text{big and positive}) \times \left(\text{something very close to } \frac{1}{2}\right).$$

We have

$$\lim_{x \to \infty} \frac{x^5 - x^2 + 1}{2x^2 + 3} = \infty.$$

Similarly, $\lim_{x \to -\infty} \frac{x^5 - x^2 + 1}{2x^2 + 3}$ wants to become

$$(\text{big and negative}) \times \left(\text{something very close to } \frac{1}{2}\right)$$

and so

$$\lim_{x \to -\infty} \frac{x^5 - x^2 + 1}{2x^2 + 3} = -\infty.$$

This rational function has no horizontal asymptotes.

In general:

Suppose a rational function f has the form

$$f(x) = \frac{ax^n + \cdots}{bx^m + \cdots}$$

with $n > m$. Then the outputs of the function either grow arbitrarily large and positive or large and negative as $x \to \pm\infty$.

All variations are possible. Consider, for instance, the long-tem behaviors of the (rational) functions given by x^2, x^3, $-x^2$, and $-x^3$.

Graphing Rational Functions

Let's study a specific example. Let's sketch the graph of the function f given by

$$f(x) = \frac{x^2 - 5x + 6}{x + 1}.$$

We can begin by exploring some basic features of the graph.

x-intercepts

The graph of the function crosses the x-axis when $y = 0$. And this occurs when $x^2 - 5x + 6 = (x - 2)(x - 3) = 0$, that is, when $x = 2$ and when $x = 3$.

y-intercept

The graph crosses the y-axis when $x = 0$. This gives $y = 6$.

Long-term behavior

We have

$$\lim_{x \to \infty} \frac{x^2 - 5x + 6}{x + 1} = \lim_{x \to \infty} x \cdot \left(\frac{1 - \frac{5}{x} + \frac{6}{x^2}}{1 + \frac{1}{x}} \right) = \text{``big} \times 1\text{''} = \infty,$$

$$\lim_{x \to -\infty} \frac{x^2 - 5x + 6}{x + 1} = \lim_{x \to -\infty} x \cdot \left(\frac{1 - \frac{5}{x} + \frac{6}{x^2}}{1 + \frac{1}{x}} \right)$$
$$= \text{``negative and big} \times 1\text{''} = -\infty.$$

So far we have identified the features of the graph given in Figure 69.

The function given by $f(x) = \frac{x^2 - 5x + 6}{x + 1}$ of course has additional features. For example, it is undefined at values for which the denominator is 0, namely, at $x = -1$. It is common practice to draw a vertical line at this location on a graph to say, in effect, don't go there! (See Figure 70.)

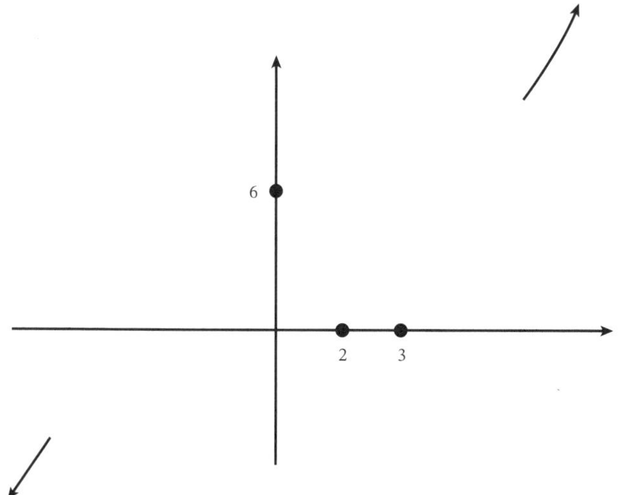

FIGURE 69

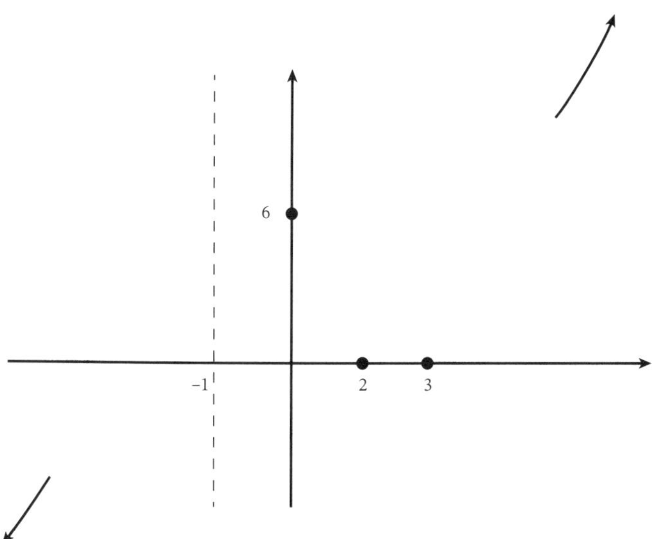

FIGURE 70

But we can be daring and ask the following:

What does the graph do as we step very close to this danger zone?

For example, let's put in a value just to the right of $x = -1$, say, $x = -0.9999$, and see what the function does for this value.

Now

$$f(x) = \frac{x^2 - 5x + 6}{x + 1} = \frac{(x - 2)(x - 3)}{x + 1}.$$

When $x = -0.9999$, the numerator is the product of two negative numbers and so is positive. The denominator is also positive, but it is a very, very small positive number. We have in effect

$$f(-0.9999) = \frac{\text{postive number}}{\text{postive number very close to 0}}.$$

This will result in a very large positive value. So we have:

Just to the right of $x = -1$ the graph is large and positive.

Now, just to the left of $x = -1$, say, at $x = -1.003$, we see that $f(x) = \frac{(x-2)(x-3)}{x+1}$ appears as

$$f(-1.003) = \frac{\text{postive number}}{\text{negative number very close to 0}}.$$

This results in a very large negative value. So we have:

Just to the left of $x = -1$ the graph is large and negative.

We can include these features in the graph. (See Figure 71.)

We have a *vertical asymptote* at $x = -1$.

We can also analyze the function just to the left and right of the zeros $x = 2$ and $x = 3$ to check that the function crosses the x-axis at these values.

These are the only broad features we can analyze without stronger techniques from calculus. Nonetheless, we can now make a pretty good guess as to the shape of the entire graph of the function. (Recall that it cannot touch the x-axis except at the points indicated.) See Figure 72.

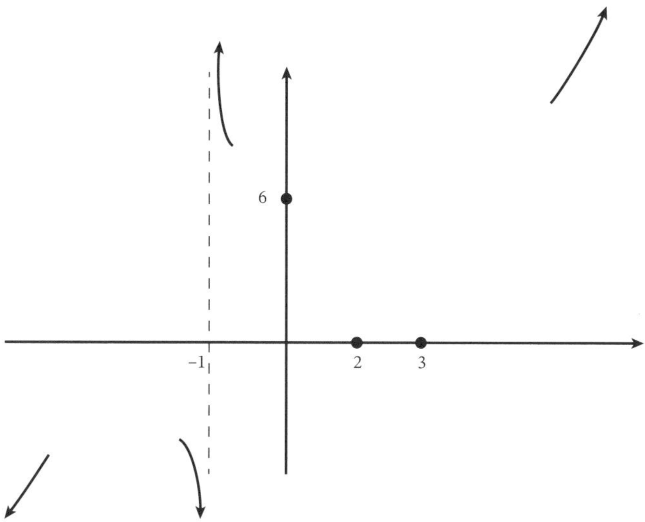

FIGURE 71

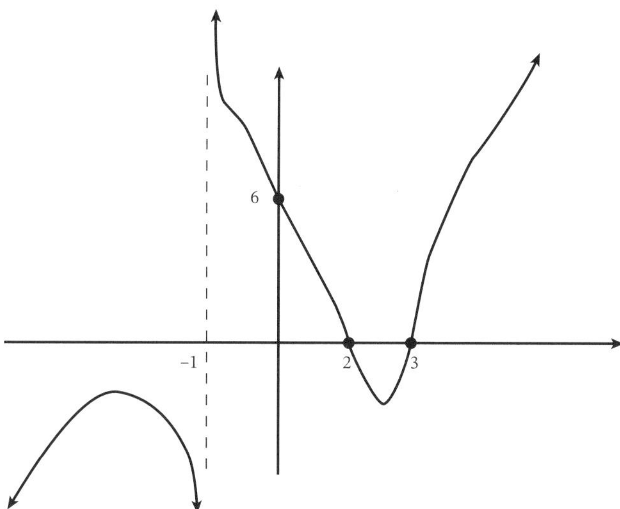

FIGURE 72

Exercise. Analyze and sketch the graph of $y = \dfrac{x^2-1}{x^2-4}$.

Solution.

x-intercepts

We have that $y = 0$ when $x^2 - 1 = 0$, that is, when $x = 1$ and $x = -1$.

y-intercept

When $x = 0$, we have $y = \frac{1}{4}$.

Long-term behavior

Because

$$\lim_{x \to \infty} \frac{x^2 - 1}{x^2 - 4} = 1,$$

$$\lim_{x \to -\infty} \frac{x^2 - 1}{x^2 - 4} = 1,$$

we have a horizontal asymptote: $y = 1$.

Vertical asymptotes

This function is undefined at $x = 2$ and $x = -2$.

To the left of $x = -2$ (say, at $x = -2.0005$) we have

$$\frac{\text{positive}}{\text{small positive}} = \text{large positive}.$$

To the right of $x = -2$ (say, at $x = -1.887$) we have

$$\frac{\text{positive}}{\text{small negative}} = \text{large negative}.$$

To the left of $x = 2$ (say, at $x = 1.99$) we have

$$\frac{\text{positive}}{\text{small negative}} = \text{large negative}.$$

To the right of $x = 2$ (say, at $x = 2.000000000009852$) we have

$$\frac{\text{positive}}{\text{small positive}} = \text{large positive}.$$

Putting this all together, this suggests a graph of the form shown in Figure 73. Again, we cannot be sure of the fine details: there may be other dips and humps.

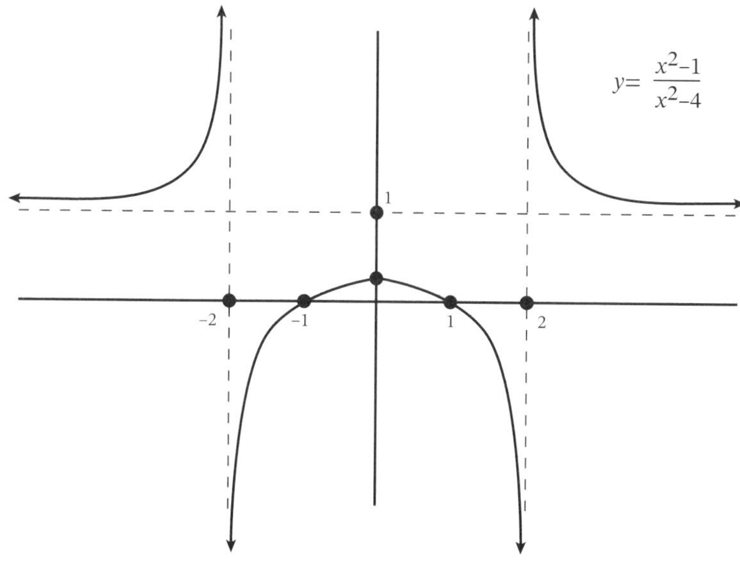

FIGURE 73

However, one can justify some additional features, for example that the function is always above the horizontal line $y = 1$ for $x > 2$:

$$
\begin{aligned}
y &= \frac{x^2 - 1}{x^2 - 4} \\
&= \frac{x^2 - 4 + 3}{x^2 - 4} \\
&= \frac{x^2 - 4}{x^2 - 4} + \frac{3}{x^2 - 4} \\
&= 1 + \frac{3}{x^2 - 4} \\
&> 1,
\end{aligned}
$$

noting that $x^2 - 4$ is positive for $x > 2$.

As a sketch of broad features, the picture we have is usually deemed sufficient. $\qquad\square$

Exercise. Sketch $y = \dfrac{x^3 + 1}{x}$.

Solution.

x-intercepts

We have that $y = 0$ when $x = -1$.

y-intercept

When $x = 0$, the function is undefined. There is no *y*-intercept.

Long-term behavior

$$\lim_{x \to \infty} \frac{x^3 + 1}{x} = \infty,$$

$$\lim_{x \to -\infty} \frac{x^3 + 1}{x} = \infty.$$

Vertical asymptotes

The function is undefined at $x = 0$.

Just to the left of $x = 0$ (say, at $x = -0.0004$) we have

$$\frac{\text{positive}}{\text{small negative}} = \text{large negative}$$

and just to the right

$$\frac{\text{positive}}{\text{small positive}} = \text{large positive}.$$

This suggests the graph of Figure 74.

Apart from a vertical asymptote at $x = 0$, this function has another type of asymptote. Notice that

$$y = \frac{x^3 + 1}{x} = x^2 + \frac{1}{x}.$$

When x is large (either large and positive or large and negative), the term $\frac{1}{x}$ is very close to 0. Thus, for extreme values of x, the function $y = \frac{x^3 + 1}{x}$ wants to approach the graph $y = x^2$. See Figure 75.

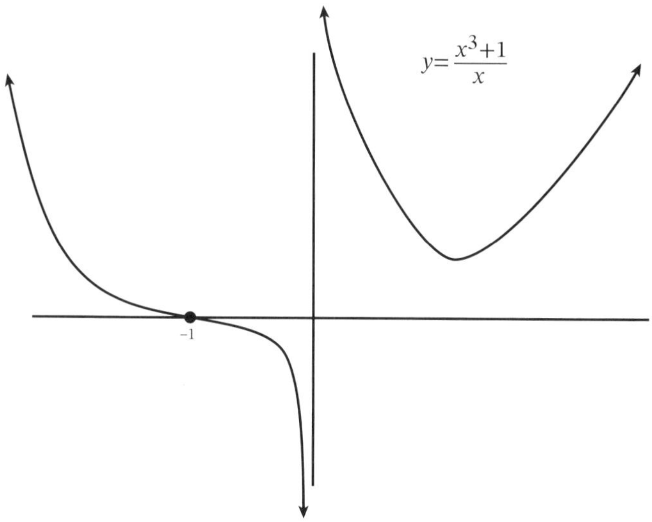

FIGURE 74

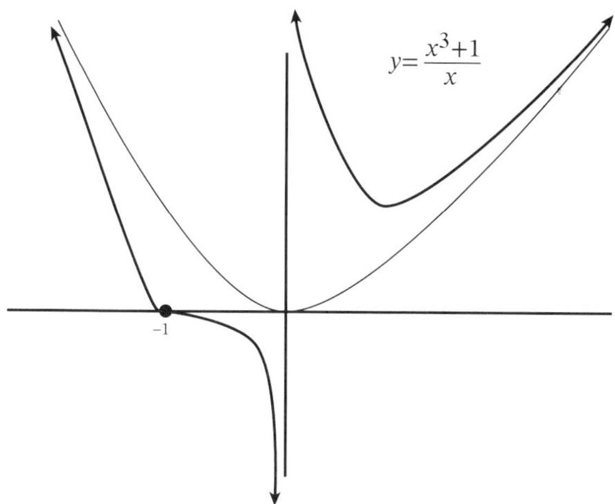

FIGURE 75

One can similarly show that $y = \dfrac{x^2+1}{x}$ has a diagonal line as an asymptote and that $y = \dfrac{x^5+32}{x^2}$ has the cubic curve $y = x^3$ as an asymptote. $\qquad\square$

Graphs of Basic Rational Functions

Some folks find it helpful to have the shapes of the graphs of $y = x^{-n} = \frac{1}{x^n}$ close in mind. The techniques described in this chapter show that they come in two basic shapes. See Figure 76.

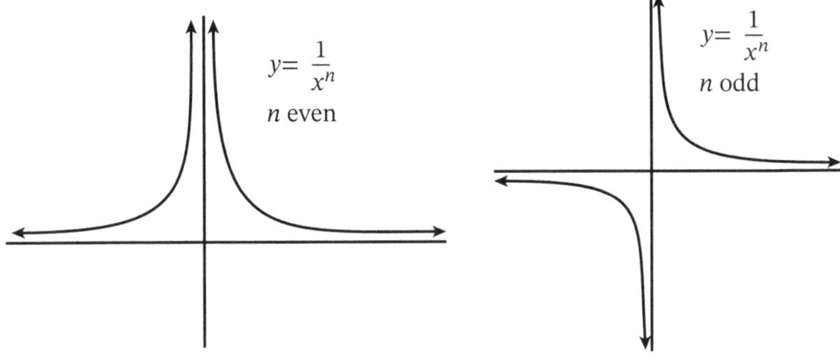

FIGURE 76

Knowing these basic graph shapes can help to swiftly sketch the graphs of other rational functions. For example, rewriting $y = \frac{x-2}{x-3}$ as $y = \frac{x-3+1}{x-3} = 1 + \frac{1}{x-3}$ shows that its graph is a translation of the graph of $y = \frac{1}{x}$. (Sketch the graph!)

The graph of $y = \frac{2+x^2}{x^2}$ is just a translation of the graph of $y = \frac{1}{x^2}$, and the graph of $y = \frac{3x^2-12x+10}{(x-2)^2}$ is just a transformation of the graph of $y = \frac{1}{x^2}$ too. (Do you see how to sketch it?)

11

Select Special Functions and Equations

In this chapter we give a swift overview of the graphs of all but one of the remaining special functions and equations that appear in a typical high school curriculum. (We refer the reader to the trigonometry guide in this series for a discussion of trigonometric functions. See the footnote on page 78.)

This section assumes familiarity with the algebra of exponents and logarithms.

Circles

The Pythagorean Theorem shows that if (x, y) is a point on a circle of radius r with center the origin, then the coordinates of that point satisfy the equation $x^2 + y^2 = r^2$. (See Figure 77.) And, conversely, if x and y are two values that make the equation $x^2 + y^2 = r^2$ true for a given positive value r, then the point (x, y) sits at the vertex of a right triangle with legs of length $|x|$ and $|y|$ and hypotenuse of length r. Thus (x, y) is a point on the circle with radius r and center the origin. Thus $x^2 + y^2 = r^2$ is the equation of a circle with center $(0, 0)$ and radius r.

The equation $(x - a)^2 + (y - b)^2 = r^2$ is essentially the same equation, except $x = a$ is now behaving like 0 for the x-values and $y = b$ is behaving like 0 for the y-values. Its graph must be a translate of a circle and so is a circle, now with center (a, b) and radius r.

Exercise. Describe the graph of the equation

$$x^2 + 6x + y^2 - 8y + 5 = 0.$$

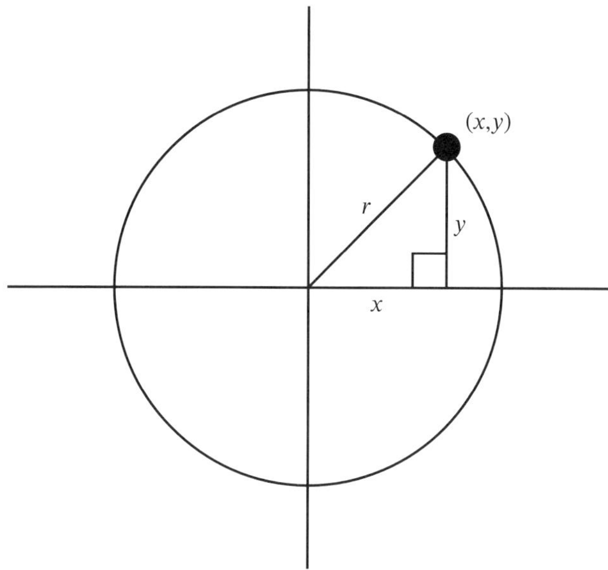

FIGURE 77

Solution. We can complete the square, twice, within the equation. We have

$$x^2 + 6x + 9 + y^2 - 8y + 16 = 20,$$
$$(x + 3)^2 + (y - 4)^2 = 20.$$

This is the equation of the circle with center $(-3, 4)$ and radius $\sqrt{20}$. $\quad\square$

Exercise. Let $A = (2, 7)$ and $B = (4, 11)$. Find the equation of the circle with $\overline{AB}$ as its diameter.

Solution. The center of the circle is the midpoint of $\overline{AB}$, which is

$$\left(\frac{2 + 4}{2}, \frac{7 + 11}{2}\right) = (3, 9).$$

The radius of the circle is $\frac{1}{2}\sqrt{(4 - 2)^2 + (11 - 7)^2} = \frac{1}{2}\sqrt{20} = \sqrt{5}$. Thus the equation of the circle is $(x - 3)^2 + (y - 9)^2 = 5$. $\quad\square$

Comment. This equation is algebraically equivalent to the equation $(x - 2)(x - 4) + (y - 7)(y - 11) = 0$. In general, if $A = (a_1, a_2)$ and

$B = (b_1, b_2)$, then the equation of the circle with $\overline{AB}$ as diameter is given by

$$(x - a_1)(x - b_1) + (y - a_2)(y - b_2) = 0.$$

Ellipses

The equation of a circle centered about the origin with radius r is given by $x^2 + y^2 = r^2$. We can rewrite this as $\frac{x^2}{r^2} + \frac{y^2}{r^2} = 1$. For some fun, let's consider an equation with a mismatch of denominators.

Example. Describe the graph of the equation $\frac{x^2}{25} + \frac{y^2}{9} = 0$.

Let's collect some data from this equation.

When $y = 0$, we have $\frac{x^2}{25} = 1$ and so the graph of the equation crosses the x-axis at -5 and 5.

When $x = 0$, we have $\frac{y^2}{9} = 1$ and so the graph of the equation crosses the y-axis at -3 and 3.

We can see that there are no data points with $x = 10$, say. (The equation $\frac{100}{25} + \frac{y^2}{9} = 0$; that is, $y^2 = -36$ has no real solutions.) Only values of x with $-5 \leq x \leq 5$ yield data points.

Playing with some of these intermediate values, identifying and plotting more data points, we see that the graph of this function is an oval-shaped curve. (See Figure 78.)

The curve in Figure 78 is called an *ellipse.* It matches the closed curve studied in classic Greek geometry that arises from slicing a cone with a plane. (This assertion requires proof.)

In general:

> The graph of the equation $\frac{x^2}{a^2} + \frac{y^2}{b^2} = 1$ is an ellipse that crosses the x axis at $\pm a$ and the y axis at $\pm b$.

If $a = b$, then the graph of the equation is a perfect circle.

Exercise. Describe the graph of $\frac{(x-78)^2}{144} + (y + 22)^2 = 1$.

Solution. This equation resembles the equation of the ellipse $\frac{x^2}{144} + \frac{y^2}{1} = 1$ centered at the origin with x-radius 12 and y-radius 1. The given equation is the same ellipse but now translated to be centered at the point $(78, -22)$. $\qquad\square$

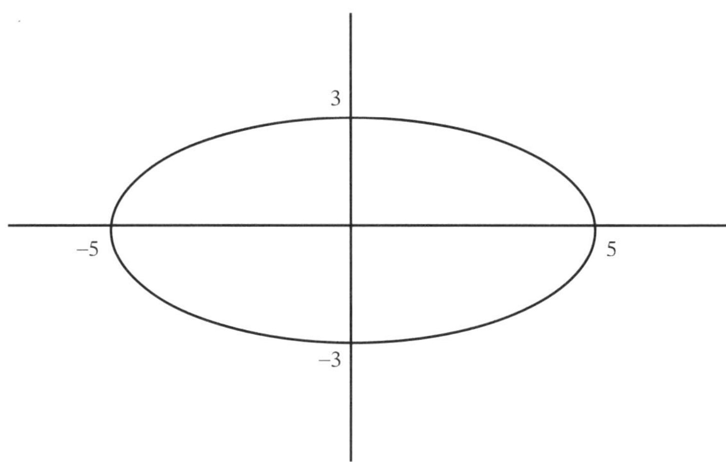

FIGURE 78

Hyperbolas

The equation of an ellipse centered at the origin is $\dfrac{x^2}{a^2} + \dfrac{y^2}{b^2} = 1$. Continuing our play, what can we say about the graph of the variant equation $\dfrac{x^2}{a^2} - \dfrac{y^2}{b^2} = 1$?

Example. Sketch a graph of $\dfrac{x^2}{4} - \dfrac{y^2}{9} = 1$.

We can identity its x-intercepts. When $y = 0$, we have $x = 2$ or -2.

The graph of this equation has no y-intercepts: when $x = 0$, we get $-\dfrac{y^2}{9} = 1$, which has no real solutions.

Unlike the equation of the ellipse, there are data points that fit this equation with x as large and positive and as large and negative as we please. Let's see if we can examine the long-term behavior of the graph of this equation.

Motivated by our work on polynomials, let's try pulling out the highest power of x that appears:

$$x^2\left(\frac{1}{4} - \frac{1}{9}\left(\frac{y}{x}\right)^2\right) = 1.$$

This can be rewritten as

$$\frac{1}{4} - \frac{1}{9}\left(\frac{y}{x}\right)^2 = \frac{1}{x^2}$$

or

$$\left(\frac{y}{x}\right)^2 = \frac{9}{4} - \frac{9}{x^2}.$$

What is this telling us?

Well, if x is a value very large in magnitude, then the ratio $\left(\frac{y}{x}\right)^2$ is very close to being the constant value $\frac{9}{4}$. So $\frac{y}{x} \approx \frac{3}{2}$ or $\frac{y}{x} \approx -\frac{3}{2}$ for very large values of x. That is, $y \approx \frac{3}{2}x$ or $y \approx -\frac{3}{2}x$. In the extreme, the graph of $\frac{x^2}{4} - \frac{y^2}{9} = 1$ thus matches a pair of crossed lines.

Some plotting of points leads to a graph that looks like Figure 79.

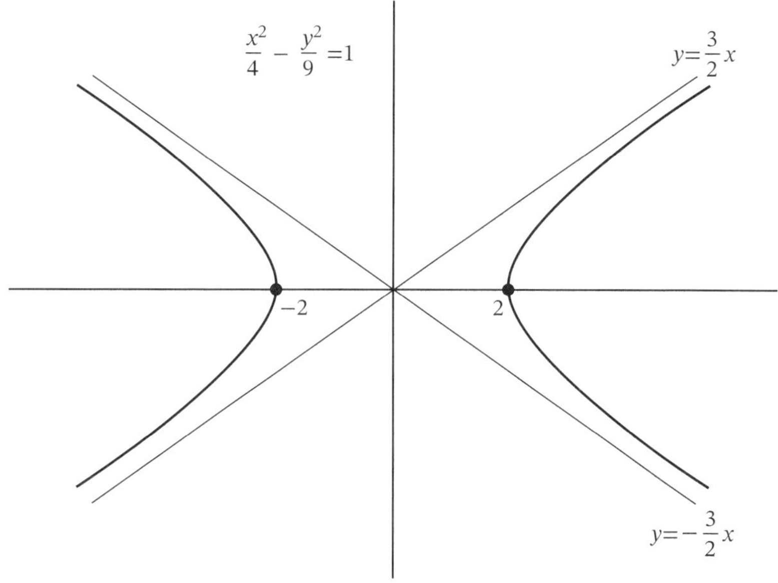

FIGURE 79

The curve in Figure 79 is called a *hyperbola*. It turns out to match one of the open curves that arises in slicing a cone with a plane as studied in classical Greek geometry. (This assertion requires proof.)

In general

> The graph of $\dfrac{x^2}{a^2} - \dfrac{y^2}{b^2} = 1$ is a hyperbola with x-intercepts $\pm a$
> and diagonal asymptotes given by the lines $y = \pm\dfrac{b}{a}x$.

Comment. Students are sometimes taught a "rectangle method" for sketching hyperbolas. The diagonals of the rectangle that students draw are just the lines $y = \pm\dfrac{b}{a}x$. I personally find it easier to just divide through by x^2 to get $\dfrac{1}{a^2} - \dfrac{1}{b^2}\cdot\dfrac{y^2}{x^2} = \dfrac{1}{x^2}$ and use the fact that $\dfrac{1}{x^2}$ is essentially 0 for large values of x. (Does this approach help you sketch a graph of $x^3 - y^3 = 1$?)

Exponential Functions

Consider the exponential equation $y = 2^x$. By plotting points we can see that the graph of this equation has the shape given by Figure 80. (Refer to `http://www.jamestanton.com/?p=379` to see how quantities such as 2^0, 2^{-1}, $2^{\frac{1}{2}}$, and $2^{\sqrt{2}}$ are properly defined.) It has the x-axis as a horizontal asymptote to the left and the y-values grow rapidly as $x \to \infty$. The graph crosses the y-axis at $y = 1$.

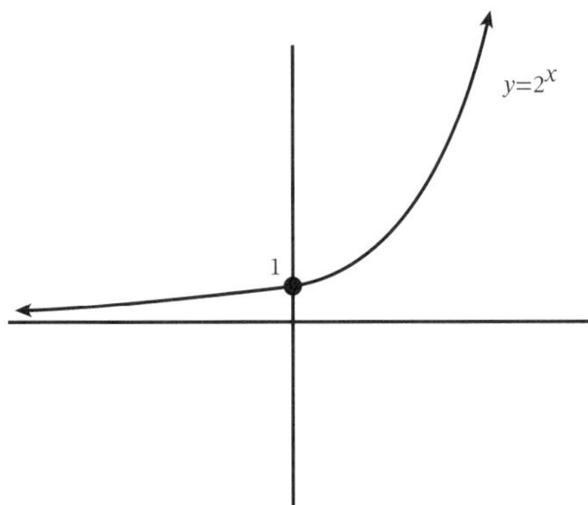

FIGURE 80

The equation $y = 3^x$ has the same features as the graph of $y = 2^x$: same y-intercept, same horizontal asymptote to the left, same basic growth to the right, but the y-values grow larger for larger and larger positive x-values more quickly than the powers of 2 (compare 3^4 with 2^4, for instance) and, for negative inputs, grow smaller more rapidly (compare 3^{-2} with 2^{-2}, for instance).

The graph of $y = 10^x$ will have the same features, but will be more pronounced in its growth and its decay. (See Figure 81.)

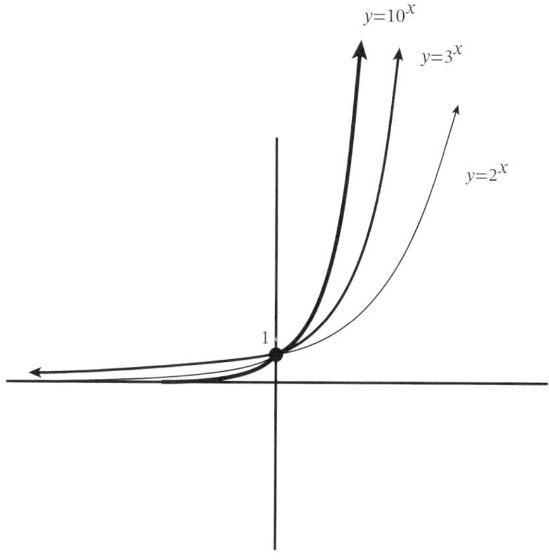

FIGURE 81

Exercise. Describe the graph of $y = \left(\frac{1}{2}\right)^x$.

Solution. One can, of course, plot data values that fit this equation to see the shape of the curve. But $y = \left(\frac{1}{2}\right)^x = 2^{-x}$. So the graph of this equation is a reflection across the y-axis of the graph of $y = 2^x$. (We are, in some sense, interchanging the roles of positive and negative x-values.) □

Exercise. Describe the graph of $y = 1^x$.

Solution. The graph of this equation is the horizontal line $y = 1$. □

A function f of the form $f(x) = a \cdot b^x$ for some constants a and b (with $a \neq 0$ and $b > 0$ and $b \neq 1$) is called an *exponential function*.

If $a > 0$ and $b > 1$, then the graph of $y = a \cdot b^x$ is similar in shape and structure to the graph of $y = 2^x$ (but the y-intercept is now $(0, a)$.)

If $a < 0$ and $b > 1$, then the graph of $y = a \cdot b^x$ is similar in shape and structure to the graph of $y = 2^x$ reflected about the horizontal axis.

If $a > 0$ and $0 < b < 1$, then the graph of $y = a \cdot b^x$ is similar in shape and structure to the graph of $y = 2^x$ reflected about the vertical axis.

If $a < 0$ and $0 < b < 1$, then the graph of $y = a \cdot b^x$ is similar in shape and structure to the graph of $y = 2^x$ reflected about both axes.

We see that an exponential function never gives an output of 0. Instead, its graph always has $y = 0$ as a horizontal asymptote either to the left or to the right.

Logarithmic Functions

The notion of a logarithm is greatly simplified if one substitutes the word *power* for the word *logarithm*. (After all, this is what they actually are!) For example, read

$$3 = \log_2 8$$

as

$$3 = \text{power}_2\, 8.$$

This says that 3 is the power of 2 that gives 8. (See `www.jamestanton.com/?p=553` for a brief history of logarithms and the reason why we don't just call them "powers".)

Consider a function that is defined by a logarithmic equation

$$y = \log_b x.$$

Reading this as "$y = \text{power}_b\, x$" we see that we have

$$x = b^y.$$

This is an equation defining an exponential function $y = b^x$, but with the roles of x and y interchanged. We see that:

A logarithm is the inverse function to an exponential function.

The graph of any logarithmic function is thus the graph of an exponential function reflected across the diagonal line. (See Figure 82.) We see from the graph that $\log_b x$, with $b > 1$, is positive for $x > 1$ and negative for $0 < x < 1$.

Exercise. Sketch the graph of $y = \log_{\frac{1}{2}} x$.

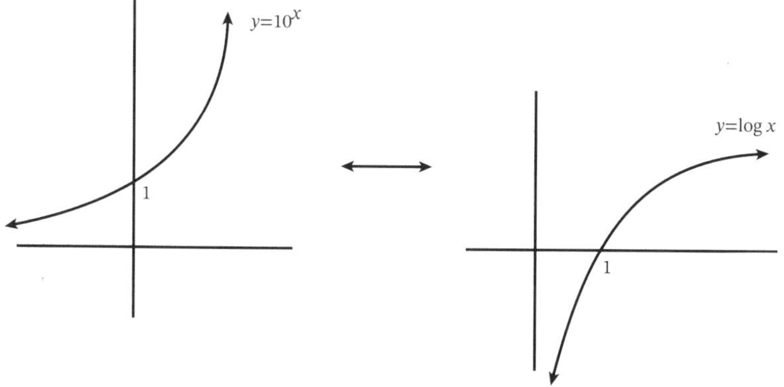

FIGURE 82

Solution. There is no embarrassment to be had in just plotting points. See Figure 83. $\square$

x	$\dfrac{1}{8}$	$\dfrac{1}{4}$	$\dfrac{1}{2}$	1	2	4	8	16
$\log_{\frac{1}{2}} x$	3	2	1	0	−1	−2	−3	−4

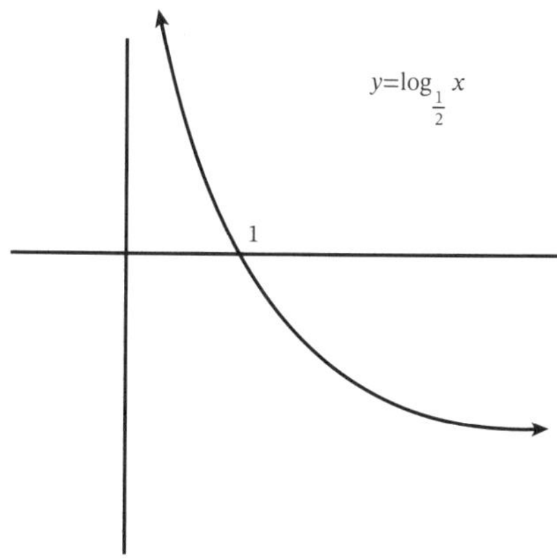

FIGURE 83

Exercise. Sketch a graph of the function $f(x) = \log_2 x$. (It will have a shape similar to the graph shown in Figure 82.)

A sketch of $y = f(4x) = \log_2(4x)$ should be the same graph compressed horizontally by a factor of 4. Sketch this graph.

The logarithm rules say this should match a sketch of $y = \log_2(x) + 2 = f(x) + 2$, the graph shifted vertically two units. Sketch this graph.

Do the two additional graphs appear to be identical?

Solution. They must be! $\qquad\qquad\qquad\qquad\qquad\qquad\qquad\square$

One can have fun with these relations. For example, the graph of $y = 4^x = 2^{(2x)}$ should match a horizontal compression, by a factor of 2, of the graph of $y = 2^x$. Does it?

If you are up for some challenges, try sketching a graph of $y = \log_x 3$ for $x > 0$. (Be careful about $x = 1$.) Using a calculator sketch a graph of $y = x^{\frac{1}{\log x}}$ for $x > 1$ and explain what you see!

MAA Featured Problem

> **(#21, AMC 12A, 2006)**
>
> Let
> $$S_1 = \left\{ (x, y) \mid \log_{10}(1 + x^2 + y^2) \leq 1 + \log_{10}(x + y) \right\}$$
> and
> $$S_2 = \left\{ (x, y) \mid \log_{10}(2 + x^2 + y^2) \leq 2 + \log_{10}(x + y) \right\}.$$
> What is the ratio of the area of S_2 to the area of S_1?
>
> (A) 98
> (B) 99
> (C) 100
> (D) 101
> (E) 102

A Personal Account of Solving This Problem

Curriculum Inspirations Strategy (`www.maa.org/ci`):

> Strategy 2: Do something!

This question is a visual nightmare! The expressions

$$S_1 = \left\{ (x,y) \mid \log_{10}(1 + x^2 + y^2) \le 1 + \log_{10}(x+y) \right\},$$
$$S_2 = \left\{ (x,y) \mid \log_{10}(2 + x^2 + y^2) \le 2 + \log_{10}(x+y) \right\}$$

look really scary. And when I read the actual question, I see it is asking about a ratio of areas. We have areas?

Deep breath.

I see that S_1 is a set of points (x,y) that satisfy some equation, oops, inequality.

Do I know what that means?

What if S_1 were friendlier, say,

$$S_1 = \left\{ (x,y) \mid 1 + x^2 + y^2 \le 1 + x + y \right\},$$

just ignoring the logarithms? That's still too hard. What about simpler still, say,

$$S_1 = \left\{ (x,y) \mid x^2 + y^2 \le 1 \right\}$$

instead? Okay, that's the set of points sitting inside a circle of radius 1. The set S_1, as originally described, also defines some region in the plane.

Okay, my job then is to make sense of the actual inequality that defines S_1 and see if I can identify what shape region it represents in the plane, and then do the same for S_2.

I really have no choice but to try to do something with the inequality

$$\log_{10}(1 + x^2 + y^2) \le 1 + \log_{10}(x+y).$$

Raise everything as a power of 10?

$$10^{\log_{10}(1+x^2+y^2)} \le 10^{1+\log_{10}(x+y)},$$
$$1 + x^2 + y^2 \le 10(x+y).$$

Oh, this looks like a circle!

$$x^2 - 10x + y^2 - 10y \le -1,$$
$$x^2 - 10x + 25 + y^2 - 10y + 25 \le 49,$$
$$(x-5)^2 + (y-5)^2 \le 49.$$

The set S_1 is the interior of a circle of radius 7 and so has area 49π.

I can see that

$$\log_{10}(2 + x^2 + y^2) \le 2 + \log_{10}(x + y)$$

gives

$$2 + x^2 + y^2 \le 100(x + y),$$

$$(x - 50)^2 + (y - 50)^2 \le 4998.$$

S_2 is the interior of a circle and has area $\pi(\sqrt{4998})^2 = 4998\pi$. The ratio of the areas is

$$\frac{4998}{49} = \frac{4900 + 98}{49} = 102.$$

Wow!

Additional Problems

60. (#16, AMC 12A, 2004) The set of real numbers x for which

$$\log_{2004}\left(\log_{2003}\left(\log_{2002}(\log_{2001} x)\right)\right)$$

is defined is $\{\, x \mid x > c \,\}$. What is the value of c?

(A) 0

(B) 2001^{2002}

(C) 2002^{2003}

(D) 2003^{2004}

(E) $2001^{2002^{2003}}$

61. (#21, AMC 12B, 2004) The graph of $2x^2 + xy + 3y^2 - 11x - 20y + 40 = 0$ is an ellipse in the first quadrant of the xy-plane. Let a and b be the maximum and minimum values of y/x over all points (x, y) on the ellipse. What is the value of $a + b$?

(A) 3

(B) $\sqrt{10}$

(C) $\dfrac{7}{2}$

(D) $\dfrac{9}{2}$

(E) $2\sqrt{14}$

12
Fitting Formulas to Data Points

Last time I checked my savings account I had a balance of $275. Checking just now I see that my balance has since grown to $825. Woohoo!

There are two ways to think about this growth. Additive thinking has me say that I gained $550 over this recent period of time: $825 = 275 + 550$. Multiplicative thinking has me say that I tripled my money: $825 = 275 \times 3$.

Let's call the time period between the two times I checked my balance one unit of time. If I believe these growth rates are constant, then additive thinking predicts I will have $275 + 550 + 550 = \$1375$ after another unit of time, and multiplicative thinking predicts a balance of $275 \times 3 \times 3 = \$2475$. (Let's hope it is multiplicative growth!) In general, after t units of time, additive thinking predicts a balance of $A(t) = 275 + 550t$ dollars and multiplicative thinking predicts a balance of $M(t) = 275 \times 3^t$ dollars.

We have just found two equations that fit the two data points $(0, 275)$ and $(1, 825)$. The first equation is a *linear model* and the second an *exponential model*. It is a standard part of the school curriculum to have students find linear and exponential equations that fit two given data points.

But what is the best way to do this work if the two data points are not as friendly? How might one go about finding a linear expression $a + bt$ and then an exponential expression $a \cdot b^t$ that fits the data $(3, 201)$ and $(17, 640)$? See Figure 84.

t	$a + bt$
3	201
17	640

t	$a \cdot b^t$
3	201
17	640

FIGURE 84

Exponential Fit

Let's do the supposedly harder one first, the hard way, with a brute force approach. (We'll learn from this process and see how we can later avoid all the hard work.)

We seek an exponential equation

$$M(t) = a \cdot b^t$$

with $M(3) = 201$ and $M(17) = 640$. We have two equations to work with:

$$201 = a \cdot b^3,$$

$$640 = a \cdot b^{17}.$$

Dividing the equations gives $b^{14} = \frac{640}{201}$ and so $b = \left(\frac{640}{201}\right)^{\frac{1}{14}}$. Now substitute this value into the first equation to see that

$$a = \frac{201}{b^3} = 201 \left(\frac{201}{640}\right)^{\frac{3}{14}}.$$

Thus

$$M(t) = 201 \left(\frac{201}{640}\right)^{\frac{3}{14}} \left(\frac{640}{201}\right)^{\frac{t}{14}}$$

is a scary looking exponential formula that does the trick. (We can simplify it a tad, I suppose.)

Let's now reflect on what we did.

What makes this problem hard is the numbers. The growth is occurring over a period of time that starts at $t = 3$ and ends at $t = 17$, a span of fourteen units of time.

Life would be so much easier if the time period was one unit of time starting at $t = 0$. See Figure 85.

$$
\begin{array}{c|c}
t & a \cdot b^t \\
\hline
0 & 201 \\
1 & 640
\end{array}
$$

FIGURE 85

Although the growth rate is a bit unfriendly (the data has grown by a factor of $\frac{640}{201}$), it is conceptually straightforward to write an exponential equation that fits this data. We see that the following works (put in $t = 0$ and $t = 1$ to check):

$$
N(t) = 201 \left(\frac{640}{201}\right)^t .
$$

But the real data isn't growing over a period of one unit of time: it grows over fourteen units of time. So we need to slow this formula down by a factor of 14. See Figure 86.

$$
\begin{array}{c|c}
t & a \cdot b^t \\
\hline
0 & 201 \\
14 & 640
\end{array}
$$

FIGURE 86

How can we do this? How can we make fourteen units of time behave, in some sense, like one unit of time? How do we make $t = 0$ and $t = 14$ behave like $t = 0$ and $t = 1$?

Some mulling and toying suggests replacing t by $\frac{t}{14}$. So let's set

$$
W(t) = 201 \left(\frac{640}{201}\right)^{\frac{t}{14}} .
$$

And putting $t = 0$ and $t = 14$ into this formula shows it is correct.

But the data we were given doesn't start at $t = 0$. It follows a span of fourteen units of time starting at $t = 3$. See Figure 87.

$$
\begin{array}{c|c}
t & a \cdot b^t \\
\hline
3 & 201 \\
17 & 640
\end{array}
$$

FIGURE 87

Can we adjust the expression we currently have so that $t = 3$ and $t = 17$ now behave like $t = 0$ and $t = 14$?

Some more mulling and toying suggests replacing t by $t - 3$. So set

$$M(t) = 201 \left(\frac{640}{201} \right)^{\frac{t-3}{14}} .$$

And inserting $t = 3$ and $t = 17$ as a check shows that this is correct.

That does it. We have a lovely exponential formula that matches the two given data points.

Challenge. Show that our formula here agrees with the brute-force answer we obtained earlier.

Although this approach seems longer and more involved, it teaches a mathematical practice: work hard to avoid hard work! We started with the simplest version of the problem and built up ideas from there.

Just Write Down the Answer!

Now that we have seen the structure of the equations, we have confidence to simply write down fitting formulas, without a lick of scratch work.

Consider, for example, the two data points $(87, 123)$ and $(1000, 14)$. Can you just see that

$$M(t) = 123 \left(\frac{14}{123} \right)^{\frac{t-87}{913}}$$

fits? (This equation represents exponential decay.)

Exercise.   (a) Write down an exponential equation that fits the data $(10, 10)$ and $(20, 20)$.
  (b) Write down an exponential equation that fits the data $(3, 46)$ and $(15, 46)$, if you can.
  (c) Write down an exponential equation that fits the data $(8, 11)$ and $(18, 0)$, if you can.

Solution. (a) $P(t) = 10 \cdot (2)^{\frac{t-10}{10}} .$

 (b) $Q(t) = 46 \cdot 1^{\frac{t-3}{12}} = 46.$

 (c) There is no exponential function of the form $f(t) = a \cdot b^t$ that fits this data. $\qquad\square$

Linear Fit

Let's repeat our "avoid the hard work" approach with additive thinking. Let's write down a linear equation for each of the three data tables in Figure 88.

t	$a + bt$
0	201
1	640

t	$a + bt$
0	201
14	640

t	$a + bt$
3	201
17	640

FIGURE 88

For the first table we can use $B(t) = 201 + 439t$.

Slowing down by a factor of 14 gives $C(t) = 201 + 439 \cdot \dfrac{t}{14}$ for the second table of values.

Making $t = 3$ behave like $t = 0$ gives $A(t) = 201 + 439 \cdot \dfrac{t-3}{14}$, which is a linear equation fitting the third data table.

Comment. The standard curriculum approach is to use the brute-force method, but with one piece of sophistication. A line connecting the two data points will have slope $\dfrac{640-201}{17-13} = \dfrac{439}{14}$, and so a linear equation that fits the data will have the form $A(t) = \dfrac{439}{14}t + b$. We can now substitute in one data point to solve for b.

Exercise. Consider the following data:

x	y
3	4.2
6	1.8

(a) Write down an exponential equation $y = a \cdot b^x$ that fits this data.

(b) Draw a table of x and $\log y$ values. Find a linear equation that fits the x and $\log y$ data.

(c) Do your equations in parts (a) and (b) match?

(d) Find an equation of the form $y = a \cdot x^b$ that fits the two original data values. (Apply a logarithm first?)

(e) Is there an equation of the form $y = x^a + b$ that fits the original data? (This is a yes/no question.)

Solution.   (a) $y = 4.2 \cdot \left(\frac{1}{3}\right)^{\frac{x-1}{5}}$.

(b) We get $\log y = -0.074x + 0.697$ (with coefficients rounded to three decimal places).

(c) The equation $y = 4.2 \cdot 3^{\frac{1-x}{5}}$ can be rewritten as

$$y = 12.6 \cdot (3^{-\frac{1}{3}})^x = 4.2 \cdot 3^{\frac{1}{5}} \cdot (3^{-\frac{1}{5}})^x \approx 5.232 \cdot 0.803^x.$$

The equation $\log y = -0.074x + 0.697$ can be rewritten as

$$y = 10^{0.697} \cdot (10^{-0.074})^x \approx 4.977 \cdot 0.843^x.$$

Are these close to being the same? With less rounding do we get a closer match? (In theory, should the two equations match precisely?)

(d) The equation $y = a \cdot x^b$ can be rewritten as $\log y = b \log x + \log a$. This is an equation of a line for data values $(\log x, \log y)$. The line though $(0, 0.623)$ and $(0.778, 0.255)$ is $\log y = -0.473 \log x + 0.623$, suggesting choosing $a = 10^{0.623} = 4.2$ and $b = -0.473$. (But there may be a concern about rounding errors to contend with.)

(e) No! (Put $x = 1$, $y = 4.2$ into this equation to see that we will need $b = 3.2$, and then put in $x = 6$, $y = 1.8$ to see there is no possible value to adopt for a.) $\square$

More than Two Data Points

Example. Find a quadratic function that fits the data table of Figure 89.

x	y
2	7
5	10
7	3

FIGURE 89

The best thing to do is to just write down the answer! Here it is:

$$p(x) = 7 \cdot \frac{(x-5)(x-7)}{(-3)(-5)} + 10 \cdot \frac{(x-2)(x-7)}{(3) \cdot (-2)} + 3\frac{(x-2)(x-5)}{(5) \cdot (2)}.$$

If we were to expand this out, we'd see that this is indeed a quadratic function. But, of course, this is not the issue in our minds right now. Where did this formula come from?

To understand this formula, start by putting in the value $x = 2$. Notice that the second and third terms are designed to vanish at $x = 2$ and so we have only to contend with the first term,

$$7 \cdot \frac{(x-5)(x-7)}{(-3)(-5)}.$$

When $x = 2$, the numerator and the denominator match (the denominator was designed to do this) so that this term becomes

$$7 \cdot 1,$$

which has the value 7. So we see now that $p(2) = 7 + 0 + 0 = 7$, as desired.

For the value $x = 5$ only the middle term

$$10 \cdot \frac{(x-2)(x-7)}{(3)\cdot(-2)}$$

survives and has value $10 \cdot \frac{(3)\cdot(-2)}{(3)\cdot(-2)} = 10$ for $x = 5$. So $p(5) = 0 + 10 + 0 = 10$, as desired.

In the same way, for the value $x = 7$ only the third term is non-vanishing and we get $P(7) = 0 + 0 + 3 \cdot \frac{5\cdot2}{5\cdot2} = 3$.

Thus the quadratic

$$p(x) = 7 \cdot \frac{(x-5)(x-7)}{(-3)(-5)} + 10 \cdot \frac{(x-2)(x-7)}{(3)\cdot(-2)} + 3\frac{(x-2)(x-5)}{(5)\cdot(2)}$$

does indeed produce the desired outputs for the given inputs. If we desires, we can simplify the expression to see that we have

$$p(x) = -\frac{9}{10}x^2 + \frac{219}{30}x - 4.$$

Despite the visual complication of the formula, we can see that its construction is relatively straightforward:

(1) Write a series of numerators that each vanish at all but one of the desired inputs.

(2) Create denominators that cancel the numerators when a specific input is entered.

(3) Use the desired output values as coefficients.

As another example, here's a quadratic that passes though the points $A = (3, 87)$, $B = (10, \pi)$, and $C = (35, \sqrt{2})$. It is

$$q(x) = 87\frac{(x - 10)(x - 35)}{(-7)(-32)} + \pi\frac{(x - 3)(x - 35)}{(7)(-28)} + \sqrt{2}\frac{(x - 3)(x - 10)}{32 \cdot 22}.$$

There is no need to set up a system of three equations in three unknowns to fit a quadratic function to three data points. (And even simplifying the quadratic expressions that appear, if needed at all, is not as tedious as it first might appear.)

Exercise. Find a quartic that fits the data of Figure 90.

$$\begin{array}{c|c} x & y \\ \hline 1 & a \\ 2 & b \\ 3 & c \\ 4 & d \\ 5 & e \end{array}$$

FIGURE 90

Solution. The polynomial

$$h(x) = a\frac{(x - 2)(x - 3)(x - 4)(x - 5)}{(-1)(-2)(-3)(-4)} + b\frac{(x - 1)(x - 3)(x - 4)(x - 5)}{(1)(-1)(-2)(-3)}$$
$$+ \cdots + e\frac{(x - 1)(x - 2)(x - 3)(x - 4)}{(4)(3)(2)(1)}$$

does the trick. $\square$

One can use this approach to write down the equation of lines. For example, a degree 1 polynomial that fits the data $(6, 2)$ and $(8, 5)$ is $y = 2\frac{(x-8)}{(-2)} + 5\frac{(x-6)}{(2)}$.

The work of this section establishes the following:

> A degree n polynomial is uniquely determined by its value on $n + 1$ distinct inputs.

Comment. See the study guide on trigonometry to review how to fit formulas to periodic phenomena. See the footnote on page 78.

Additional Problem

62. (#24, AMC 12A, 2005) Let $P(x) = (x - 1)(x - 2)(x - 3)$. For how many polynomials $Q(x)$ does there exist a polynomial of degree 3 such that $P(Q(x)) = P(x) \cdot R(x)$?

(A) 19

(B) 22

(C) 24

(D) 27

(E) 32

Part II
Solutions

Solutions

Here are previously published solutions to the competition problems as they appear at `www.edfinity.com` or in one of the MAA's published texts on the AMC competitions. (Go to `https://bookstore.ams.org/maa-press-browse.`)

Warning. Each solution presented here is fast paced and to the point, simply working through the mathematics of the task at hand to get the job done. These solutions are written by a variety of authors.

For an account of the problem-solving practices behind each solution guiding the mathematical steps presented—along with discussion on its connections to the Common Core State Standards and further, deeper, queries and possible explorations—see the Curriculum Bursts at `www.maa.org/ci`.

Chapter 1

1. (E) Suppose $N = 10a + b$. Then $10a + b = ab + (a + b)$. It follows that $9a = ab$, which implies that $b = 9$, since $a \neq 0$. Notice that the numbers $19, 29, 39, \ldots, 99$ all meet the required condition.

2. (E) Let $y = \clubsuit(x)$. Since each digit of x is at most 9, we have $y \leq 18$. Thus if $\clubsuit(y) = 3$, then $y = 3$ or $y = 12$. The three values of x for which $\clubsuit(x) = 3$ are 12, 21, and 30, and the seven values of x for which $\clubsuit(x) = 12$ are 39, 48, 57, 66, 75, 84, and 93. There are ten values in all.

Chapter 2

3. (B) Let the sequence be denoted $a, ar, ar^2, ar^3, \ldots$, with $ar = 2$ and $ar^3 = 6$. Then $r^2 = 3$ and $r = \sqrt{3}$ or $r = -\sqrt{3}$. Therefore $a = \frac{2\sqrt{3}}{3}$ or $a = -\frac{2\sqrt{3}}{3}$.

4. (C) Let D be the difference between consecutive terms of the sequence. Then $a - c = 2D, b = c - D, d = c + D$, and $e = c + 2D$,

so

$$a + b + c + d + e = (c - 2D) + (c - D) + c + (c + D) + (c + 2D) = 5c.$$

Thus $5c = 30$, so $c = 6$.

To see that the values of the other terms cannot be found, note that the sequences 4, 5, 6, 7, 8 and 10, 8, 6, 4, 2 both satisfy the given conditions.

5. (C) Let $d = a_2 - a_1$. Then $a_{k+100} = a_k + 100d$ and

$$a_{101} + a_{102} + \cdots + a_{200} = (a_1 + 100d) + (a_2 + 100d)$$
$$+ \cdots + (a_{100} + 100d)$$
$$= a_1 + a_2 + \cdots + a_{100} + 10000d.$$

Thus

$$200 = 100 + 10000d \quad \text{and} \quad d = \frac{100}{10000} = 0.01.$$

6. (D) We have

$$a_2 = a_{1+1} = a_1 + a_1 + 1 \cdot 1 = 1 + 1 + 1 = 3,$$
$$a_3 = a_{2+1} = a_2 + a_1 + 2 \cdot 1 = 3 + 1 + 2 = 6,$$
$$a_6 = a_{3+3} = a_3 + a_3 + 3 \cdot 3 = 6 + 6 + 9 = 21,$$

and

$$a_{12} = a_{6+6} = a_6 + a_6 + 6 \cdot 6 = 21 + 21 + 36 = 78.$$

7. (D) Note that

$$a_{2^1} = a_2 \ = a_{2 \cdot 1} = 1 \cdot a_1 = 2^0 \cdot 2^0 \ = 2^0,$$
$$a_{2^2} = a_4 \ = a_{2 \cdot 2} = 2 \cdot a_2 = 2^1 \cdot 2^0 \ = 2^1,$$
$$a_{2^3} = a_8 \ = a_{2 \cdot 4} = 4 \cdot a_4 = 2^2 \cdot 2^1 \ = 2^{1+2},$$
$$a_{2^4} = a_{16} = a_{2 \cdot 8} = 8 \cdot a_8 = 2^3 \cdot 2^{1+2} = 2^{1+2+3},$$

and, in general,

$$a_{2^n} = 2^{1+2+3+\cdots+(n-1)}.$$

Because $1 + 2 + 3 + \cdots + (n - 1) = \frac{1}{2}n(n - 1)$, we have

$$a_{2^{100}} = 2^{(100)(99)/2} = 2^{4950}.$$

8. (C) Let a_k be the kth term of the sequence. For $k \geq 3$, we have $a_{k+1} = a_{k-2} + a_{k-1} - a_k$, so $a_{k+1} - a_{k-1} = -(a_k - a_{k-2})$. Because the sequence begins

$$2001, 2002, 2003, 2000, 2005, 1998, \ldots$$

it follows that the odd-numbered terms and the even-numbered terms each form arithmetic progressions with common differences of 2 and -2, respectively. The 2004th term of the original sequence is the 1002nd term of the sequence 2002, 2000, 1998, …, and that term is $2002 + 1001(-2) = 0$.

9. (E) The sequence begins 2005, 133, 55, 250, 133, …. Thus after the initial term 2005, the sequence repeats the cycle 133, 55, 250. Because $2005 = 1 + 3 \cdot 668$, the 2005th term is the same as the last term of the repeating cycle, 250.

10. (E) Note that the first several terms of the sequence are

$$2, 3, \frac{3}{2}, \frac{1}{2}, \frac{1}{3}, \frac{2}{3}, 2, 3, \dots$$

so the sequence consists of a repeating cycle of 6 terms. Since $2006 = 2 + 334 \cdot 6$, we have $a_{2006} = a_2 = 3$.

11. (C) The sum of a set of real numbers is the product of the mean and the number of real numbers, and the median of a set of consecutive integers is the same as the mean. So the median must be $7^5/49 = 7^3$.

12. (C) We have $b = a + r$, $c = a + 2r$, and $d = a + 3r$, where r is a positive real number. Also $b^2 = ad$ yields $(a + r)^2 = a(a + 3r)$, or $r^2 = ar$. It follows that $r = a$ and $d = a + 3a = 4a$. Hence $\frac{a}{d} = \frac{1}{4}$.

13. (E) The terms involving only odd powers of r form the geometric series $ar + ar^3 + ar^5 + \cdots$. Thus

$$7 = a + ar + ar^2 + \cdots = \frac{a}{1 - r},$$

and

$$3 = ar + ar^3 + ar^5 + \cdots$$
$$= \frac{ar}{1 - r^2}$$
$$= \frac{a}{1 - r} \cdot \frac{r}{1 - r}$$
$$= \frac{7r}{1 - r}.$$

Therefore $r = 3/4$. It follows that $a/(1/4) = 7$, so $a = 7/4$ and

$$a + r = \frac{7}{4} + \frac{3}{4} = \frac{5}{2}.$$

14. (A) The terms of the arithmetic progression are $9, 9 + d$, and $9 + 2d$ for some real number d. The terms of the geometric progression are 9, $11 + d$, and $29 + 2d$. Therefore $(11 + d)^2 = 9(29 + 2d)$ so $d^2 + 4d - 140 = 0$. Thus $d = 10$ or $d = -14$. The corresponding geometric progressions are $9, 21, 49$ and $9, -3, 1$, so the smallest possible value for the third term of the geometric progression is 1.

15. (B) Note that beginning with the third term of the sequence, each term depends only on the previous two. Writing out more terms of the sequence yields

$$4, 7, 1, 8, 9, 7, 6, 3, 9, 2, 1, 3, 4, 7, 1, \ldots .$$

Therefore the sequence is periodic with period 12. The sequence repeats itself, starting with the 13th term. Since $S_{12} = 60$, $S_{12k} = 60k$ for all positive integers k. The largest k for which $S_{12k} \leq 10{,}000$ is

$$k = \left\lfloor \frac{10{,}000}{60} \right\rfloor = 166,$$

and $S_{12 \cdot 166} = 60 \cdot 166 = 9960$. To have $S_n > 10{,}000$, we need to add enough additional terms for the sum to exceed 40. This can be done by adding the next 7 terms of the sequence, since their sum is 42. Thus the smallest value of n is $12 \cdot 166 + 7 = 1999$.

16. (A) Since $2002 = 11 \cdot 13 \cdot 14$, we have

$$a_n = \begin{cases} 11, & \text{if } n = 13 \cdot 14 \cdot i, \text{ where } i = 1, 2, \ldots, 10, \\ 13, & \text{if } n = 14 \cdot 11 \cdot j, \text{ where } j = 1, 2, \ldots, 12, \\ 14, & \text{if } n = 11 \cdot 13 \cdot k, \text{ where } k = 1, 2, \ldots, 13, \\ 0, & \text{otherwise.} \end{cases}$$

Hence $\sum_{n=1}^{2001} a_n = 11 \cdot 10 + 13 \cdot 12 + 14 \cdot 13 = 448$.

17. (D) If a, b, and c are three consecutive terms of such a sequence, then $ac - 1 = b$, so $c = (1 + b)/a$. Applying this rule recursively and simplifying yields

$$\ldots, a, b, \frac{1 + b}{a}, \frac{1 + a + b}{ab}, \frac{1 + a}{b}, a, b, \ldots .$$

This shows that at most five different terms can appear in such a sequence. Moreover, the value a is determined once the value 2000 is assigned to b and the value 2001 is assigned to another of the five terms. Thus there are four such sequences that contain 2001 as a term, namely

$$2001, 2000, 1, \frac{1}{1000}, \frac{1001}{1000}, 2001, \dots,$$

$$1, 2000, 2001, \frac{1001}{1000}, \frac{1}{1000}, 1, \dots,$$

$$\frac{2001}{4001999}, 2000, 4001999, 2001, \frac{2002}{4001999}, \frac{2001}{4001999}, \dots,$$

and

$$4001999, 2000, \frac{2001}{4001999}, \frac{2002}{4001999}, 2001, 4001999, \dots.$$

The four values of x are 2001, 1, $\frac{2001}{4001999}$, and 4001999.

18. (B) The condition $a_{n+2} = |a_{n+1} - a_n|$ implies that a_n and a_{n+3} have the same parity for all $n \geq 1$. Because a_{2006} is odd, a_2 is also odd. Because $a_{2006} = 1$ and a_n is a multiple of $\gcd(a_1, a_2)$ for all n, it follows that $1 = \gcd(a_1, a_2) = \gcd(3^3 \cdot 37, a_2)$. There are 499 odd integers in the interval $[1, 998]$, of which 166 are multiples of 3, 13 are multiples of 37, and 4 are multiples of $3 \cdot 37 = 111$. By the Inclusion-Exclusion Principle, the number of possible values of a_2 cannot exceed $499 - 166 - 13 + 4 = 324$.

To see that there are actually 324 possibilities, note that for $n \geq 3$, $a_n < \max(a_{n-2}, a_{n-1})$ whenever a_{n-2} and a_{n-1} are both positive. Thus $a_N = 0$ for some $N \leq 1999$. If $\gcd(a_1, a_2) = 1$, then $a_{N-2} = a_{N-1} = 1$, and for $n > N$ the sequence cycles through the values 1, 1, 0. If in addition a_2 is odd, then a_{3k+2} is odd for $k \geq 1$, so $a_{2006} = 1$.

19. (E) Note that

$$z_{n+1} = \frac{iz_n}{\bar{z}_n} = \frac{iz_n^2}{z_n\bar{z}_n} = \frac{iz_n^2}{|z_n|^2}.$$

Since $|z_0| = 1$, the sequence satisfies

$$z_1 = iz_0^2,$$
$$z_2 = iz_1^2 = i(iz_0^2)^2 = -iz_0^4,$$

and, in general, when $k \geq 2$,

$$z_k = -iz_0^{2^k}.$$

Hence z_0 satisfies the equation $1 = -iz_0^{(2^{2005})}$, so $z_0^{(2^{2005})} = i$. Because every nonzero complex number has n distinct nth roots, this equation has 2^{2005} solutions. So there are 2^{2005} possible values for z_0.

Chapter 3

20. (A) The value of $3x - 1$ is 5 when $x = 2$. Thus $f(5) = f(3 \cdot 2 - 1) = 2^2 + 2 + 1 = 7$.

21. (C) Note that $f(600) = f\left(500 \cdot \frac{6}{5}\right) = \frac{f(500)}{6/5} = \frac{3}{6/5} = \frac{5}{2}$.

Note. $f(x) = \frac{1500}{x}$ is the unique function satisfying the given conditons.

22. (C) From the definition of f,

$$f(x + iy) = i(x - iy) = y + ix$$

for all real numbers x and y, so the numbers that satisfy $f(z) = z$ are the numbers of the form $x + ix$. The set of all such numbers is a line through the origin in the complex plane. The set of all numbers that satisfy $|z| = 5$ is a circle centered at the origin of the complex plane. The numbers satisfying both equations correspond to the points of intersection of the line and circle, of which there are two.

23. (E) The conditions on f imply that both $x = f(x) + f\left(\frac{1}{x}\right)$ and $\frac{1}{x} = f\left(\frac{1}{x}\right) + f\left(\frac{1}{1/x}\right) = f\left(\frac{1}{x}\right) + f(x)$. Thus if x is in the domain of f, then $x = 1/x$, so $x = \pm 1$.

The conditions are satisfied if and only if $f(1) = 1/2$ and $f(-1) = -1/2$.

Chapter 4

24. (D) If $f(x) = mx + b$, then

$$12 = f(6) - f(2) = 6m + b - (2m + b) = 4m,$$

so $m = 3$. Hence

$$f(12) - f(2) = 12m + b - (2m + b) = 10m = 30.$$

25. (A) Since $f\left(f^{-1}(x)\right) = x$, it follows that $a(bx+a)+b = x$, so $ab = 1$ and $a^2 + b = 0$. Hence $a = b = -1$, so $a + b = -2$.

Chapter 5

26. (E) Substitute $(2, 3)$ and $(4, 3)$ into the equation to give

$$3 = 4 + 2b + c \quad \text{and} \quad 3 = 16 + 4b + c.$$

Subtracting corresponding terms in these equations gives $0 = 12 + 2b$. So

$$b = -6 \quad \text{and} \quad c = 3 - 4 - 2(-6) = 11.$$

27. (C) The equation $(x + y)^2 = x^2 + y^2$ is equivalent to $x^2 + 2xy + y^2 = x^2 + y^2$, which reduces to $xy = 0$. Thus the graph of the equation consists of the two lines that are the coordinate axes.

28. (E) Substituting $x = 1$ and $y = 2$ into the equations gives $1 = \dfrac{2}{4} + a$ and $2 = \dfrac{1}{4} + b$.

It follows that

$$a + b = \left(1 - \frac{2}{4}\right) + \left(2 - \frac{1}{4}\right) = 3 - \frac{3}{4} = \frac{9}{4}.$$

29. (B) We have

$$0 = f(-1) = -a + b - c + d \quad \text{and} \quad 0 = f(1) = a + b + c + d,$$

so $b + d = 0$. Also $d = f(0) = 2$, so $b = -2$.

Chapter 6

30. (D) The given equation implies that either

$$x - 1 = x - 2 \quad \text{or} \quad x - 1 = -(x - 2).$$

The first equation has no solution, and the solution to the second equation is $x = 3/2$.

31. (D) The graph is symmetric with respect to both coordinate axes, and in the first quadrant it coincides with the graph of the line $3x + 4y = 12$. Therefore the region is a rhombus, and the area is

$$\text{Area} = 4\left(\frac{1}{2}(4 \cdot 3)\right) = 24.$$

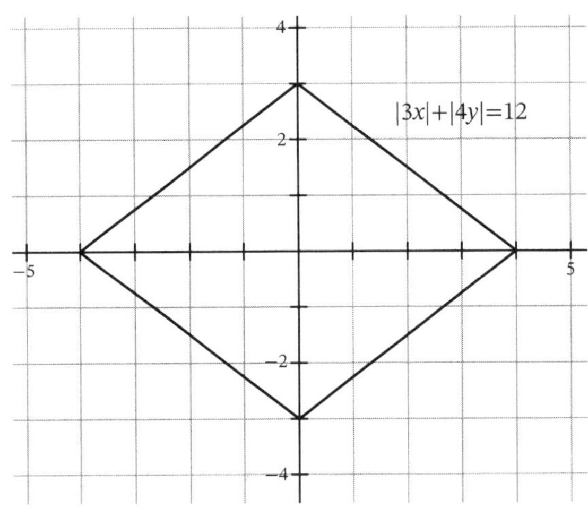

(This diagram is a redrawing of the original figure.)

Chapter 7

32. (B) The y-intercept of the line is between 0 and 1, so $0 < b < 1$. The slope is between -1 and 0, so $-1 < m < 0$. Thus $-1 < mb < 0$.

33. (A) The two lines have equations

$$y - 15 = 3(x - 10) \quad \text{and} \quad y - 15 = 5(x - 10).$$

The x-intercepts, obtained by setting $y = 0$ in the respective equations, are 5 and 7. The distance between the points $(5, 0)$ and $(7, 0)$ is 2.

34. (D) The equation $f(f(x)) = 6$ implies that $f(x) = -2$ or $f(x) = 1$. The horizontal line $y = -2$ intersects the graph twice, so $f(x) = -2$ has two solutions. Similarly, $f(x) = 1$ has four solutions, so there are six solutions of $f(f(x)) = 6$.

35. (C) Since the system has exactly one solution, the graphs of the two equations must intersect at exactly one point. If $x < a$, the equation $y = |x - a| + |x - b| + |x - c|$ is equivalent to $y = -3x + (a + b + c)$.

By similar calculations we obtain

$$y = \begin{cases} -3x + (a + b + c) & \text{if } x < a, \\ -x + (-a + b + c) & \text{if } a \leq x < b, \\ x + (-a - b + c) & \text{if } b \leq x < c, \\ 3x + (-a - b - c) & \text{if } c \leq x. \end{cases}$$

Thus the graph consists of four lines with slopes $-3, -1, 1$, and 3, and it has corners at $(a, b - c - 2a), (b, c - a)$, and $(c, 2c - a - b)$.

On the other hand, the graph of $2x + y = 2003$ is a line whose slope is -2. If the graphs intersect at exactly one point, that point must be $(a, b + c - 2a)$. Therefore

$$2003 = 2a + (b + c - 2a) = b + c.$$

Since $b < c$, the minimum value of c is 1002.

Chapter 8

36. (A) Factor to get

$$0 = (2x + 3)(x - 4) + (2x + 3)(x - 6) = (2x + 3)(2x - 10).$$

The two roots are $-3/2$ and 5, which sum to $7/2$.

37. (C) The given conditions imply that

$$x^2 + ax + b = (x - a)(x - b) = x^2 - (a + b)x + ab,$$

so

$$a + b = -a \quad \text{and} \quad ab = b.$$

Since $b \neq 0$, the second equation implies that $a = 1$. The first equation gives $b = -2$, so $(a, b) = (1, -2)$.

38. (B) Since $0 = 2x^2 + 3x - 5 = (2x + 5)(x - 1)$, we have $d = -\dfrac{5}{2}$ and $e = 1$. So $(d - 1)(e - 1) = 0$.

39. (A) The quadratic formula yields

$$x = \frac{-(a + 8) \pm \sqrt{(a + 8)^2 - 4 \cdot 4 \cdot 9}}{2 \cdot 4}.$$

The equation has only one solution precisely when the value of the discriminant is 0, that is, when

$$(a + 8)^2 - 144 = 0.$$

This implies that $a + 8 = \pm 12$. So $a = -20$ or $a = 4$, and the sum is -16.

40. (D) Let r_1 and r_2 be the roots of $x^2 + px + m = 0$. Then $0 = x^2 + px + m = (x - r_1)(x - r_2)$, so $m = r_1 r_2$ and $p = -(r_1 + r_2)$. Since the roots of $x^2 + mx + n = 0$ are $2r_1$ and $2r_2$, we also have $0 = x^2 + mx + n = (x - 2r_1)(x - 2r_2)$, so $n = 4r_1 r_2$ and $m = -2(r_1 + r_2)$. Thus

$$n = 4m, \quad p = \frac{1}{2}m, \quad \text{and} \quad \frac{n}{p} = \frac{4m}{\frac{1}{2}m} = 8.$$

41. (D) Because a and b are roots of $x^2 - mx + 2 = 0$, we have $x^2 - mx + 2 = (x - a)(x - b)$ and $ab = 2$.

In a similar manner, the constant term of $x^2 - px + q$ is the product of $a + (1/b)$ and $b + (1/a)$, so

$$q = \left(a + \frac{1}{b}\right)\left(b + \frac{1}{a}\right) = ab + 1 + 1 + \frac{1}{ab} = \frac{9}{2}.$$

42. (B) Let $a = 2003/2004$. The given equation is equivalent to

$$ax^2 + x + 1 = 0.$$

If the roots of this equation are denoted r and s, then

$$rs = \frac{1}{a} \quad \text{and} \quad r + s = -\frac{1}{a},$$

so $\dfrac{1}{r} + \dfrac{1}{s} = \dfrac{r+s}{rs} = -1.$

43. (B) Let p and q be two primes that are roots of $x^2 - 63x + k = 0$. Then

$$x^2 - 63x + k = (x - p)(x - q) = x^2 - (p + q)x + p \cdot q,$$

so $p + q = 63$ and $p \cdot q = k$. Since 63 is odd, one of the primes must be 2 and the other 61. Thus there is exactly one possible value for k, namely $k = p \cdot q = 2 \cdot 61 = 122$.

44. (C) The vertex of the parabola is $(0, a^2)$. The line passes through the vertex if and only if $a^2 = 0 + a$. There are two solutions to this equation, namely $a = 0$ and $a = 1$.

45. (D) A parabola with the given equation and with vertex (p, p) must have equation $y = a(x - p)^2 + p$. Because the y-intercept is $(0, -p)$ and $p \neq 0$, it follows that $a = -2/p$. Thus

$$y = -\frac{2}{p}(x^2 - 2px + p^2) + p = -\frac{2}{p}x^2 + 4x - p,$$

so $b = 4$.

46. (E) The equation of the first parabola can be written in the form

$$y = a(x - h)^2 + k = ax^2 - 2axh + ah^2 + k,$$

and the equation of the second (having the same shape and vertex, but opening in the opposite direction) can be written in the form

$$y = -a(x - h)^2 + k = -ax^2 + 2axh - ah^2 + k.$$

Hence

$$a + b + c + d + e + f$$
$$= a + (-2ah) + (ah^2 + k) + (-a) + (2ah) + (-ah^2 + k)$$
$$= 2k.$$

47. (C) A number x differs by 1 from its reciprocal if and only if $x - 1 = 1/x$ or $x + 1 = 1/x$. These equations are equivalent to $x^2 - x - 1 = 0$ and $x^2 + x - 1 = 0$. Solving these by the quadratic formula yields the positive solutions

$$\frac{1 + \sqrt{5}}{2} \quad \text{and} \quad \frac{-1 + \sqrt{5}}{2},$$

which are reciprocals of each other. The sum of the two numbers is $\sqrt{5}$.

48. (D) The original parabola has equation $y = a(x - h)^2 + k$ for some a, h, and k, with $a \neq 0$. The reflected parabola has equation $y = -a(x - h)^2 - k$. The translated parabolas have equations

$$f(x) = a(x - h \pm 5)^2 + k \quad \text{and} \quad g(x) = -a(x - h \mp 5)^2 - k,$$

so

$$(f + g)(x) = \pm 20a(x - h).$$

Since $a \neq 0$, the graph is a nonhorizontal line.

49. (E) Let $B = (a, b)$ and $A = (-a, -b)$. Then

$$4a^2 + 7a - 1 = b \quad \text{and} \quad 4a^2 - 7a - 1 = -b.$$

Subtracting gives $b = 7a$, so $4a^2 + 7a - 1 = 7a$. Thus

$$a^2 = \frac{1}{4} \quad \text{and} \quad b^2 = (7a)^2 = \frac{49}{4},$$

so

$$AB = 2\sqrt{a^2 + b^2} = 2\sqrt{\frac{50}{4}} = 5\sqrt{2}.$$

50. (A) The product of the zeros of f is c/a, and the sum of the zeros is $-b/a$. Because these two numbers are equal, $c = -b$, and the sum of the coefficients is $a + b + c = a$, which is the coefficient of x^2. To see that none of the other choices is correct, let $f(x) = -2x^2 - 4x + 4$. The zeros of f are $-1 \pm \sqrt{3}$, so the sum of the zeros, the product of the zeros, and the sum of the coefficients are all -2. However, the coefficient of x is -4, the y-intercept is 4, the x-intercepts are $-1 \pm \sqrt{3}$, and the mean of the x-intercepts is -1.

51. (D) Let l be the directrix of the parabola, and let C and D be the projections of F and B onto l, respectively. For any point in the parabola, its distance to F and to l is the same. Because V and B are on the parabola, it follows that $p = FV = VC$ and $2p = FC = BD = FB$. By the Pythagorean Theorem, $VB = \sqrt{FV^2 + FB^2} = \sqrt{5}p$, and thus $\cos(\angle FVB) = \dfrac{FV}{VB} = \dfrac{p}{\sqrt{5}p} = \dfrac{\sqrt{5}}{5}$. Because $\angle AVB = 2(\angle FVB)$, it follows that

$$\cos(\angle AVB) = 2\cos^2(\angle FVB) - 1$$
$$= 2\left(\frac{\sqrt{5}}{5}\right)^2 - 1$$
$$= \frac{2}{5} - 1$$
$$= -\frac{3}{5}.$$

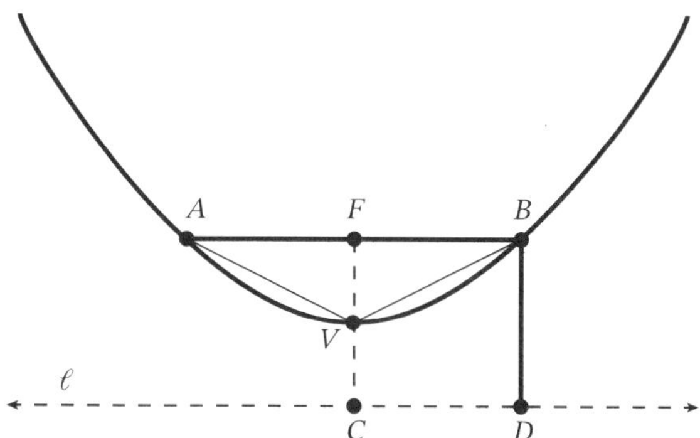

52. (C) The domain of f is $\{\, x \mid ax^2 + bx \geq 0 \,\}$. If $a = 0$, then for every positive value of b, the domain and range of f are each equal to the interval $[0, \infty)$, so 0 is a possible value of a.

If $a \neq 0$, the graph of $y = ax^2 + bx$ is a parabola with x-intercepts at $x = 0$ and $x = -b/a$. If $a > 0$, the domain of f is $(-\infty, -b/a] \cup [0, \infty)$, but the range of f cannot contain negative numbers. If $a < 0$, the domain of f is $[0, -b/a]$. The maximum value of f occurs halfway between the x-intercepts, at $x = -b/2a$, and

$$f\left(-\frac{b}{2a}\right) = \sqrt{a\left(\frac{b^2}{4a^2}\right) + b\left(-\frac{b}{2a}\right)}$$

$$= \frac{b}{2\sqrt{-a}}.$$

Hence the range of f is $[0, b/2\sqrt{-a}]$. For the domain and range to be equal, we must have

$$-\frac{b}{a} = \frac{b}{2\sqrt{-a}}$$

so

$$2\sqrt{-a} = a.$$

The only solution is $a = -4$. Thus there are two possible values of a, and they are $a = 0$ and $a = -4$.

Chapter 9

53. (B) If $n \geq 4$, then

$$n^2 - 3n + 2 = (n - 1)(n - 2)$$

is the product of two integers greater than 1 and thus is not prime. For $n = 1, 2$, and 3, respectively,

$$(1 - 1)(1 - 2) = 0,$$
$$(2 - 1)(2 - 2) = 0,$$
$$(3 - 1)(3 - 2) = 2.$$

Therefore $n^2 - 3n + 2$ is prime only when $n = 3$.

54. (A) Let r_1, r_2, and r_3 be the roots. Then

$$5 = \log_2 r_1 + \log_2 r_2 + \log_2 r_3 = \log_2 r_1 r_2 r_3,$$

so

$$r_1 r_2 r_3 = 2^5 = 32.$$

Since

$$8x^3 + 4ax^2 + 2bx + a = 8(x - r_1)(x - r_2)(x - r_3),$$

it follows that

$$a = -8r_1 r_2 r_3 = -256.$$

55. (D) Because $f(x)$ has real coefficients and $2i$ and $2 + i$ are zeros, so are their conjugates $-2i$ and $2 - i$. Therefore

$$f(x) = (x + 2i)(x - 2i)(x - (2 + i))(x - (2 - i))$$
$$= (x^2 + 4)(x^2 - 4x + 5)$$
$$= x^4 - 4x^3 + 9x^2 - 16x + 20.$$

Hence $a + b + c + d = -4 + 9 - 16 + 20 = 9$.

56. (D) Since $P(0) = 0$, we have $e = 0$ and

$$P(x) = x(x^4 + ax^3 + bx^2 + cx + d).$$

Suppose that the four remaining x-intercepts are at p, q, r, and s. Then

$$x^4 + ax^3 + bx^2 + cx + d = (x - p)(x - q)(x - r)(x - s),$$

and $d = pqrs \neq 0$.

Any of the other constants could be 0. For example, consider

$$P_1(x) = x^5 - 5x^3 + 4x = x(x+2)(x+1)(x-1)(x-2)$$

and

$$P_2(x) = x^5 - 5x^4 + 20x^2 - 16x = x(x+2)(x-1)(x-2)(x-4).$$

57. (A) If r and s are the integer zeros, the polynomial can be written in the form

$$P(x) = (x-r)(x-s)(x^2 + \alpha x + \beta).$$

The coefficients of x^3 and x^2 are, respectively, $\alpha - (r+s)$ and $\beta - \alpha(r+s) + rs$. Because r and s are integers, so are α and β. Applying the quadratic formula gives the remaining zeros as

$$\frac{1}{2}\left(-\alpha \pm \sqrt{\alpha^2 - 4\beta}\right) = -\frac{\alpha}{2} \pm i\frac{\sqrt{4\beta - \alpha^2}}{2}.$$

Answer choices (A), (B), (C), and (E) require that $\alpha = -1$, which implies that the imaginary parts of the remaining zeros have the form $\pm\sqrt{4\beta - 1}/2$. This is true only for choice (A).

Note that choice (D) is not possible since this choice requires $\alpha = -2$. This produces an imaginary part of the form $\sqrt{\beta - 1}$, which cannot be $\frac{1}{2}$.

58. (C) Let α denote the root that is an integer. Because the coefficient of x^3 is 1, there can be no other rational roots, so the two other zeros must be $\frac{\alpha}{2} \pm r$ for some irrational number r. The polynomial is then

$$(x - \alpha)\left[x - \left(\frac{\alpha}{2} + r\right)\right]\left[x - \left(\frac{\alpha}{2} - r\right)\right]$$
$$= x^3 - 2\alpha x^2 + \left(\frac{5}{4}\alpha^2 - r^2\right)x - \alpha\left(\frac{1}{4}\alpha^2 - r^2\right).$$

Therefore $\alpha = 1002$ and the polynomial is

$$x^3 - 2004x^2 + (5(501)^2 - r^2)x - 1002((501)^2 - r^2).$$

All coefficients are integers if and only if r^2 is an integer, and the zeros are positive and distinct if and only if $1 \leq r^2 \leq 501^2 - 1 = 251{,}000$. Because r cannot be an integer, there are $251{,}000 - 500 = 250{,}000$ possible values of n.

59. (B) The sum of the coefficients of P and the sum of the coefficients of Q will be equal, so $P(1) = Q(1)$. The only answer choice with an intersection at $x = 1$ is (B). (The polynomials in graph (B) are $P(x) = 2x^4 - 3x^2 - 3x - 4$ and $Q(x) = -2x^4 - 2x^3 - 2x - 2$.)

Chapter 11

60. (B) The expression is defined if and only if

$$\log_{2003}\left(\log_{2002}\left(\log_{2001} x\right)\right) > 0,$$

that is, if and only if

$$\log_{2002}(\log_{2001} x) > 2003^0 = 1.$$

This inequality in turn is satisfied if and only if

$$\log_{2001} x > 2002,$$

that is, if and only if $x > 2001^{2002}$.

61. (C) A line $y = mx$ intersects the ellipse in 0, 1, or 2 points. The intersection consists of exactly one point if and only if $m = a$ or $m = b$. Thus a and b are the values of m for which the system

$$2x^2 + xy + 3y^2 - 11x - 20y + 40 = 0,$$

$$y = mx$$

has exactly one solution. Substituting mx for y in the first equation gives

$$2x^2 + mx^2 + 3m^2x^2 - 11x - 20mx + 40 = 0,$$

or, by rearranging the terms,

$$(3m^2 + m + 2)x^2 - (20m + 11)x + 40 = 0.$$

The discriminant of this equation is

$$(20m + 11)^2 - 4 \cdot 40 \cdot (3m^2 + m + 2) = -80m^2 + 280m - 199,$$

which must be 0 if $m = a$ or $m = b$. Thus $a + b$ is the sum of the roots of the equation $-80m^2 + 280m - 199 = 0$, which is $\dfrac{280}{80} = \dfrac{7}{2}$.

Chapter 12

62. (B) The polynomial $P(x) \cdot R(x)$ has degree 6, so $Q(x)$ must have degree 2. Therefore Q is uniquely determined by the ordered triple $(Q(1), Q(2), Q(3))$. When $x = 1, 2$, or 3, we have $0 = P(x) \cdot R(x) = P(Q(x))$. It follows that $(Q(1), Q(2), Q(3))$ is one of the 27 ordered triples (i, j, k), where i, j, and k can be chosen from the set $\{1, 2, 3\}$.

However, the choices $(1, 1, 1)$, $(2, 2, 2)$, $(3, 3, 3)$, $(1, 2, 3)$, and $(3, 2, 1)$ lead to polynomials $Q(x)$ defined by $Q(x) = 1$, 2, 3, x, and $4 - x$, respectively, all of which have degree less than 2. The other 22 choices for $(Q(1), Q(2), Q(3))$ yield noncollinear points, so in each case $Q(x)$ is a quadratic polynomial.

Part III
Appendices

Appendix I
Ten Problem-Solving Strategies

Here, in brief, are the ten problem-solving strategies that apply particularly well to solving competition mathematics problems. Please go to `www.maa.org/ci` for full explanations of these strategies and further practice problems.

Remember that these strategies come after conducting the first, and most important, step in problem solving:

> **Step 1**: Read the question, have an emotional reaction to it, take a deep breath, and then reread the question.

Here are the strategies.

1. Engage in Successful Flailing

Often one can identify which topic a challenge belongs to—this question is about right triangles, or this question is about repeating decimals—but still have no clue as to how to start on the challenge. The thing to do then is to engage in direct flailing.

To do this, think about everything you know about right triangles or about repeating decimals. Read the question out loud and then describe it again in different words. Draw a picture. Try an example with actual numbers. Mark something on the diagram. And so on. Do everything you can think of that is relevant to the content at hand. In doing so, a step forward with the problem often emerges.

2. Do Something!

Innovation in research and business is not easy. Many times one is stymied and not even able to conceive of any next step to take. This can happen in mathematics problem solving too.

Perhaps the most powerful of problem-solving techniques, in mathematics and in life, is to simply DO SOMETHING, no matter how irrelevant or unhelpful it may seem. Unblock the emotional or intellectual impasse by writing something—almost anything—on the page. Turn the diagram upside down and shade in a feature that now stands out to you. Underline some words in the question statement. Just do something!

3. Engage in Wishful Thinking

As I tell my students, if there is something in life you want, make it happen. (And then be willing to handle the consequences of your actions with care and grace!) If you wish, for example, that "+4" appeared on the left side of an equation, then make it happen by adding a 4 to the left side (with the consequence of adding a 4 to the right side as well).

A beautiful problem-solving technique is to simply change the problem to what you wish it could be.

4. Draw a Picture

The upper school mathematics curriculum tends to drill down to analytic techniques and visualization is put to the side. But visual thinking can unlock deep insight. Don't underestimate the power of drawing a picture for the problem. (For example, perhaps the best way to "prove" that $1 + 2 + 3 + 4 + 5 + 4 + 3 + 2 + 1 = 5^2$ is to draw in the diagonals of a 5×5 array of dots.)

5. Solve a Smaller Version of the Same Problem

A large, complex task can be made comprehensible by examining a smaller, analogous version.

For example, if there are areas to find, can I use symmetry to my advantage and work out the area of just one piece? If there is a list, instead of finding the hundredth number, can I just find the third to get a feel for things? If I make a first move, is the rest of the game just a smaller version of the original game? Is the count of things I don't want easier to compute than the count of things I do want? And so on.

6. Eliminate Incorrect Choices

Eliminating what cannot be right helps determine that which must be correct.

Must the answer be even or odd?

Must the answer be large or small?

Should the answer involve π?

Should the graph be straight or concave?

Should the numbers increase or decrease?

With how many 0's should the answer end?

And so on.

7. Perseverance Is Key

What impression of the mathematical pursuit do we give students? Answers are preknown (they are at the back of the book or are in the teacher's mind), all can be accomplished in a fixed amount of time (quizzes and exams are usually timed), and the goal is to follow a pre-set intellectual path (as dictated by a curriculum). Open research and problem-solving tasks, on the other hand, usually possess none of these features. Persistence and perseverance are key skills absolutely vital for any success in original intellectual endeavors.

8. Second-Guess the Author

If a problem feels staged, then use that to your advantage!

The number 131 mentioned is prime. Coincidence?

The number $203 = 7 \times 29$ only has one two-digit factor. Is that helpful?

Why are we rolling the dice first and then tossing the coin? Is that order important?

Why are we focusing on factors of 2 and 5? Is it because $2 \times 5 = 10$? Hmm. $2^{10} = 1024$ is very close to 1000.

Both equations involve $\sqrt{2x}$. Is that a coincidence?

Why the arc of a circle? Is that because we want the ship to stay the same distance from a certain point?

Why a parabola? Do we need equal distances of some kind?

9. Avoid Hard Work

No one enjoys hard computation or a tedious grind through formulas and equations. Brute-force work should be undertaken only as a last resort. Do what a mathematician does—think long and hard to devise a creative, elegant approach that avoids hard work.

Do I really need to work out $2^{21} - 1$ to see if it factors?

See `www.maa.org/ci` for some examples of this strategy in action.

10. Go to Extremes

It is fun to be quirky and push ideas to the edge. Taking the parameters of a problem to an extreme can give insight to the workings of the situation. And such insight can often illuminate a path for success.

If the escalator had zero velocity (that is, it wasn't moving) how many steps would I have to climb? What if it was moving really fast?

What if the number had very few factors? What if the number was prime?

What if everyone had the same age?

What if the point P was on top of point Q? What if the point was very far away?

What if the cake's temperature was a billion degrees?

What if x was really close to 0?

What if the circle was so big that it is was practically a straight line?

Appendix II
Connections to the Common Core State Standards: Practice Standards and Content Standards

Here is an outline of how the content of this guide connects with the Common Core State Standards.

The very first Standard for Mathematical Practice asks—requires!—that we educators pay explicit attention to teaching problem solving:

MP1 Make sense of problems and persevere in solving them.

And one can argue that several, if not all, of the remaining seven Standards for Mathematical Practice can play prominent roles in supporting this first standard. For example, when solving a problem, students will likely be engaging in the activities of these standards too:

MP2 Reason abstractly and quantitatively.

MP3 Construct viable arguments and critique the reasoning of others.

MP7 Look for and make use of structure.

This, and all the guides in the *Clever Studying* series, align directly with the Standards for Mathematical Practice.

This guide also directly addresses the following Content Standards. (Relevant chapter numbers appear after each standard.)

HSF.IF.A.1 Understand that a function from one set (called the domain) to another set (called the range) assigns to each element of the domain exactly one element of the range. If f is a function and x is an element of its domain, then $f(x)$ denotes the output of f corresponding to the input x. The graph of f is the graph of the equation $y = f(x)$. (Chapter 1)

HSF.IF.A.2 Use function notation, evaluate functions for inputs in their domains, and interpret statements that use function notation in terms of a context. (Chapter 2)

HSF.IF.A.3 Recognize that sequences are functions, sometimes defined recursively, whose domain is a subset of the integers. *For example, the Fibonacci sequence is defined recursively by $f(0) = f(1) = 1$, $f(n + 1) = f(n) + f(n - 1)$ for $n \geq 1$.* (Chapter 2)

HSF.IF.B.4 For a function that models a relationship between two quantities, interpret key features of graphs and tables in terms of the quantities, and sketch graphs showing key features given a verbal description of the relationship. *Key features include: intercepts; intervals where the function is increasing, decreasing, positive, or negative; relative maximums and minimums; symmetries; end behavior; and periodicity.* (Chapter 5)

HSF.IF.B.5 Relate the domain of a function to its graph and, where applicable, to the quantitative relationship it describes. *For example, if the function $h(n)$ gives the number of person-hours it takes to assemble n engines in a factory, then the positive integers would be an appropriate domain for the function.* (Chapters 1 and 5)

HSF.IF.B.6 Calculate and interpret the average rate of change of a function (presented symbolically or as a table) over a specified interval. Estimate the rate of change from a graph. (Chapter 7)

HSS.IF.B.6.A Fit a function to the data; use functions fitted to data to solve problems in the context of the data. Use given functions or choose a function suggested by the context. Emphasize linear, quadratic, and exponential models. (Chapter 12)

HSF.IF.C.7.A Graph linear and quadratic functions and show intercepts, maxima, and minima. (Chapter 8)

HSF.IF.C.7.C Graph polynomial functions, identifying zeros when suitable factorizations are available and showing end behavior. (Chapter 9)

HSF.IF.C.7.D (+) Graph rational functions, identifying zeros and asymptotes when suitable factorizations are available and showing end behavior. (Chapter 10)

HSF.IF.C.7.E Graph exponential and logarithmic functions, showing intercepts and end behavior, and graph trigonometric functions, showing period, midline, and amplitude. (Chapter 11)

HSF.IF.C.8 Write a function defined by an expression in different but equivalent forms to reveal and explain different properties of the function. (Chapter 3)

HSF.IF.C.8.A Use the process of factoring and completing the square in a quadratic function to show zeros, extreme values, and symmetry of the graph, and interpret these in terms of a context. (Chapter 8)

HSF.IF.C.9 Compare properties of two functions each represented in a different way (algebraically, graphically, numerically in tables, or by verbal descriptions). *For example, given a graph of one quadratic function and an algebraic expression for another, say which has the larger maximum.* (Chapter 8)

HSF.BF.A.1.C (+) Compose functions. *For example, if $T(y)$ is the temperature in the atmosphere as a function of height and $h(t)$ is the height of a weather balloon as a function of time, then $T(h(t))$ is the temperature at the location of the weather balloon as a function of time.* (Chapters 1 and 4)

HSF.BF.A.2 Write arithmetic and geometric sequences both recursively and with an explicit formula, use them to model situations, and translate between the two forms. (Chapter 2)

HSF.BF.B.3 Identify the effect on the graph of replacing $f(x)$ by $f(x) + k$, $kf(x)$, $f(kx)$, and $f(x + k)$ for specific values of k (both positive and negative); find the value of k given the graphs. Experiment with cases and illustrate an explanation of the effects on the graph using technology. Include recognizing even and odd functions from their graphs and algebraic expressions for them. (Chapter 6)

HSF.BF.B.4.A Solve an equation of the form $f(x) = c$ for a simple function f that has an inverse and write an expression for the inverse. *For example, $f(x) = 2x^3$ or $f(x) = (x + 1)/(x - 1)$ for $x \neq 1$.* (Chapter 4)

HSF.BF.B.4.B (+) Verify by composition that one function is the inverse of another. (Chapter 4)

HSF.BF.B.4.D (+) Produce an invertible function from a noninvertible function by restricting the domain. (Chapter 4)

HSA.APR.A.1 Understand that polynomials form a system analogous to the integers; namely, they are closed under the operations of addition, subtraction, and multiplication. Add, subtract, and multiply polynomials. (Chapter 9)

HSA.APR.B.2 Know and apply the Remainder Theorem: for a polynomial $p(x)$ and a number a, the remainder on division by $x - a$ is $p(a)$, so $p(a) = 0$ if and only if $(x - a)$ is a factor of $p(x)$. (Chapter 9)

HSA.APR.B.3 Identify zeros of polynomials when suitable factorizations are available, and use the zeros to construct a rough graph of the function defined by the polynomial. (Chapter 9)

HSG.GPE.A.1 Derive the equation of a circle of given center and radius using the Pythagorean Theorem; complete the square to find the center and radius of a circle given by an equation. (Chapter 11)